Advances in New Green Road Materials and Applied Technologies

Advances in New Green Road Materials and Applied Technologies

Editors

Chaohui Wang
Dawei Wang
Kai Liu
Qian Chen

MDPI

Basel • Beijing • Wuhan • Barcelona • Belgrade • Novi Sad • Cluj • Manchester

Editors

Chaohui Wang
School of Highway
Chang'an University
Xi'an
China

Dawei Wang
School of Transportation
Science and Engineering
Harbin Institute of Technology
Harbin
China

Kai Liu
School of Automobile and
Transportation Engineering
Hefei University of Technology
Hefei
China

Qian Chen
School of Highway
Chang'an University
Xi'an
China

Editorial Office
MDPI
St. Alban-Anlage 66
4052 Basel, Switzerland

This is a reprint of articles from the Special Issue published online in the open access journal *Sustainability* (ISSN 2071-1050) (available at: www.mdpi.com/journal/sustainability/special_issues/Green_Road_Materials_Applied_Technologies).

For citation purposes, cite each article independently as indicated on the article page online and as indicated below:

Lastname, A.A.; Lastname, B.B. Article Title. *Journal Name* **Year**, *Volume Number*, Page Range.

ISBN 978-3-7258-0114-5 (Hbk)
ISBN 978-3-7258-0113-8 (PDF)
doi.org/10.3390/books978-3-7258-0113-8

Contents

Preface

Dear colleagues,

With the continuous development of society and the gradual improvement in road engineering constructions, humans have increasingly urgent requirements for road service functions, green construction and safety guarantees. Since entering the 21st century, the emergence of new functional materials and the development of interdisciplinary approaches have provided strong support for the design and construction of all kinds of green roads. At present, building environmentally friendly green and low-carbon roads, expanding road service functions, improving road ecological benefits and further improving road environments have become important challenges for road workers and researchers, as well as frontier directions for the development of the road engineering discipline.

To this end, scholars around the world have carried out a great deal of in-depth research on green road materials and applied technologies, and a number of important innovative results have been achieved, which is of great significance for promoting road service performance and reducing the consumption of road maintenance resources.

This Special Issue gathers recent research that advances knowledge about new green road materials and applied technologies. The supplement to these studies will guide the development of functional road materials and promote the design optimization and technological innovation of green roads. This Special Issue focuses on academic, empirical and case studies on the following research topics:

- Polyvinyl alcohol fiber-engineered cementitious composites;
- Asphalt-based waterproof bonding layers;
- Basalt fiber microwave-absorbing shielding layers;
- Asphalt rubber stress-absorbing membrane interlayers;
- Reclaimed asphalt mixtures;
- Polyurethane mixtures;
- Nano TiO_2 photocatalytic asphalt pavements;
- High-modulus asphalt mixtures;
- Steel slag/asphalt mixtures.

Chaohui Wang, Dawei Wang, Kai Liu, and Qian Chen
Editors

sustainability

MDPI

Article

Durability Performance of PVA Fiber Cement-Stabilized Macadam

Songyuan Tan [1], Chaohui Wang [1,*], Qi Zheng [2], Feng Chen [3] and Yunjie Huang [1]

1 School of Highway, Chang'an University, Xi'an 710064, China
2 Hebei Polytechnic Institute, Shijiazhuang 050020, China
3 Shanxi Transportation Technology Consulting Co., Ltd., Xi'an 710068, China
* Correspondence: wchh0205@chd.edu.cn

Abstract: To further improve the durability of cement-stabilized macadam and guarantee the use quality and sustainability of a semi-rigid base, the current study was carried out. With the help of a dry shrinkage test, temperature shrinkage test, freeze–thaw bending test, and fatigue test, the effect of incorporating PVA fiber on the deformation characteristics of cement-stabilized macadam was analyzed, and the changes in low-temperature residual toughness of the mixture before and after modification were compared. The low-temperature toughness of PVA fiber cement-stabilized macadam was evaluated with the help of the standard toughness evaluation method. The fatigue life prediction equation of PVA fiber cement-stabilized macadam was established based on the Weibull distribution. The results showed that PVA fiber can effectively improve the deformation characteristics, low-temperature toughness, and fatigue performance of cement-stabilized macadam. The low-temperature residual flexural tensile strength and low-temperature bearing capacity were increased by 10.3% and 55.3%, respectively. The residual toughness indices were increased by 58.6%, 88.1%, and 98.3% and the residual strength index was increased by more than 100%. The fatigue life was improved by 178~368% under different stress intensity ratios. The fatigue life values obeyed the two-parameter Weibull distribution, and the correlation between the fatigue life prediction equation and the measured data was significant. The fatigue life prediction error was between 0.03 and 4.9% under different stress intensity ratios.

Keywords: road material; PVA fiber cement-stabilized macadam; durability performance; low-temperature toughness

Citation: Tan, S.; Wang, C.; Zheng, Q.; Chen, F.; Huang, Y. Durability Performance of PVA Fiber Cement-Stabilized Macadam. *Sustainability* **2022**, *14*, 16953. https://doi.org/10.3390/su142416953

Academic Editor: Antonio D'Andrea

Received: 13 November 2022
Accepted: 14 December 2022
Published: 17 December 2022

1. Introduction

Cement-stabilized macadam has been widely used in pavement bases, due to its high bearing capacity and stiffness. Its anti-cracking, anti-deformation and durability directly affect the service life of the road [1–3]. Long-term research and engineering practice have found that cement-stabilized macadam is prone to cracking due to the lack of resistance to dry shrinkage, temperature shrinkage, and fatigue resistance, resulting in reflection cracks in the asphalt surface layer, which significantly affect the quality of road use. In this regard, current research usually incorporates fibers in cement-stabilized materials to enhance their durability, thus ensuring the sustainability of the semi-rigid base. The most commonly used fibers include steel fiber, carbon fiber, polypropylene fiber, and polyvinyl alcohol fiber. Polyvinyl alcohol fiber (PVA) has the advantages of high tensile strength, low elongation, and good bonding with cement-like materials, and its incorporation into cement-stabilized macadam can effectively improve early cracking problems [4–7]. This has increasingly led scholars to try to use PVA in cement-based materials.

In recent years, scholars have studied the durability performance of PVA fiber cement-based composites. Alam et al. investigated the fatigue performance of PVA fiber cement-based composites [8]. Zhang et al. and Kim et al. simulated the durability performance of PVA fiber cement concrete under a chloride salt environment [9,10]. Huang et al. and Zhang et al. analyzed the flexural toughness of PVA fiber cement-based composites using fracture toughness, instability toughness, and fracture energy [11–13]. Liu et al. evaluated the durability performance of PVA fiber cement-based composites using an adaptive network-based fuzzy inference system and constructed a structural framework for durability evaluation [14]. Li studied the effect of PVA fiber dosing and length on the shrinkage characteristics and fatigue performance of cement-stabilized macadam [15]. Yu et al. analyzed the fatigue performance of PVA fiber cement-stabilized macadam and established a cumulative fatigue life damage model [16]. Zhao et al. prepared PVA fiber cement-stabilized macadam with different PVA fiber lengths, doping amounts, and different cement contents and studied their shrinkage characteristics [17]. In summary, the current research on the durability of PVA fiber cement-based materials is mainly focused on PVA fiber cement concrete. Research on the durability of PVA fiber cement-stabilized macadam is relatively weak, mainly involving shrinkage characteristics and fatigue performance. The lack of in-depth systematic research on the bending toughness of PVA fiber cement-stabilized macadam and its evaluation, which is an important factor to ensure the durability of cement-stabilized macadam, and its performance decay directly affects the development and evolution of road performance. Therefore, it is necessary to carry out research related to the bending toughness of PVA fiber cement-stabilized macadam.

Based on this, this paper uses surface-treated PVA fibers in cement-stabilized macadam, deeply analyzes the effects of PVA fiber on the deformation characteristics and low-temperature toughness of cement-stabilized macadam and systematically evaluates the low-temperature residual toughness of PVA fiber cement-stabilized macadam. On this basis, this work further analyzes the durability performance of PVA fiber cement-stabilized macadam in combination with fatigue performance tests, and establishes the fatigue life prediction equation of PVA fiber cement-stabilized macadam based on Weibull distribution, so as to provide a useful reference for the durability improvement of PVA fiber cement-stabilized macadam.

2. Materials and Methods

2.1. Materials

PVA fiber with a length of 12 mm and diameter of 40 μm was selected. The fiber dose was 1 kg/m^3, and the control group with a fiber dose of 0% was used for comparison. The fiber surface modifier used epibromohydrin. P.O 42.5 ordinary silicate cement was selected as the cement, and limestone was used as the main aggregate, and the relevant technical indices of the cement and aggregate all met the JTG/T F20-2015 [18]. Referring to the JTG/T F20-2015 [18], the gradation of the mix was C-B-3 and the grade curve is shown in Figure 1. The maximum dry density of cement-stabilized macadam is 2.28 g/cm^3, and the optimum moisture content is 4.75% (Table 1).

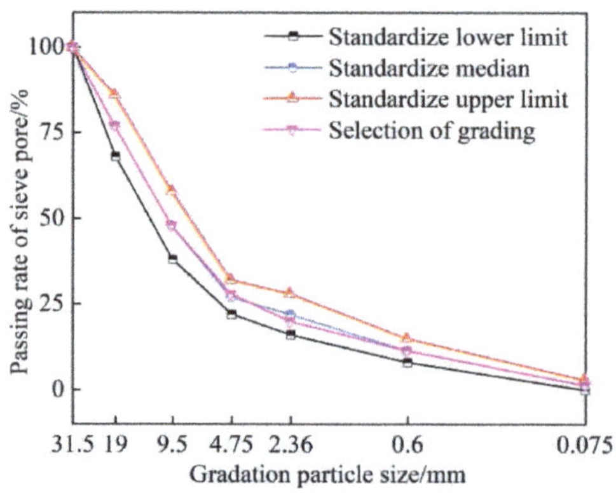

Figure 1. The cement-stabilized macadam grading curve.

Table 1. Physical and mechanical properties of fiber parameters.

Length/mm	Diameter/μm	Aspect Ratio	Density/(g·m^{-3})	Modulus of Elasticity/GPa	Tensile Strength/MPa	Elongation/%
12	40 ± 5	0.3	1.30	≥35	≥1500	≤7

2.2. Fiber Surface Modification

The following chemical methods [19] were used to modify the surface of PVA fibers: ① the appropriate amount of PVA fiber and an appropriate amount of water were added into a beaker, stirring at high speed to form a suspension. ② Epibromohydrin solution was added dropwise and the solution pH was adjusted to be greater than 11 by adding the appropriate amount of NaOH. The reaction was left at room temperature for 5 h. ③ The fibers were filtered, washed, and dried.

2.3. Testing Methods

(1) Dry shrinkage test

The dry shrinkage test was carried out according to JTG E51-2009 [20]. The water loss rate, dry shrinkage, dry shrinkage strain, dry shrinkage coefficient, and total dry shrinkage coefficient were calculated according to Equations (1)–(5), respectively.

$$\omega_i = (m_i - m_{i+1})/m_p \tag{1}$$

$$\delta_i = (\sum_{j=1}^{4} X_{i,j} - \sum_{j=1}^{4} X_{i+1,j})/2 \tag{2}$$

$$\varepsilon_i = \frac{\delta_i}{l} \tag{3}$$

$$\alpha_{di} = \frac{\varepsilon_i}{\omega_i} \tag{4}$$

$$\alpha_d = \frac{\sum \varepsilon_i}{\sum \omega_i} \tag{5}$$

where ω_i is the ith water loss (%), δ_i is the ith observed shrinkage (mm), ε_i is the ith shrinkage strain (%), α_{di} is the ith dry shrinkage coefficient (%), m_i is the ith standard specimen weighing mass (g), $X_{i,j}$ is the reading of the jth micrometer at the ith test (mm), l is the length of the standard specimen (mm) and m_p is the standard specimen after drying the constant amount (g).

(2) Temperature shrinkage test

The temperature shrinkage test was carried out with reference to JTG E51-2009 [20]. The temperature shrinkage strain and temperature shrinkage coefficient were calculated according to Equations (6) and (7).

$$\varepsilon_i = \frac{l_i - l_{i+1}}{l_0} \tag{6}$$

$$\alpha_t = \frac{\varepsilon_i}{t_i - t_{i+1}} \tag{7}$$

where l_i is the micrometer reading and average value of the ith temperature interval (mm), t_i is the ith temperature interval set by the temperature control program (°C), l_0 is the initial length of the specimen (mm) and ε_i is the average shrinkage strain of the ith temperature (%).

(3) Freeze–thaw bending test

The freeze–thaw bending test was carried out with reference to T0858-2009 in JTG E51-2009 [20]. After the freeze–thaw cycle, the specimen was placed on the UTM for the three-point bending test. The flexural tensile strength and residual flexural tensile strength were calculated according to Equations (8) and (9).

$$R_S = \frac{PL}{hb^2} \quad (8)$$

where R_S is the flexural tensile strength of the specimen (MPa), P is the damage limit load (MPa), L is the span (mm), b is the width of the specimen (mm), and h is the height of the specimen (mm).

$$CYQ = \frac{Q_{DR}}{Q_C} \times 100 \quad (9)$$

where CYQ is the residual flexural tensile strength of the specimen after n freeze–thaw cycles (%), Q_{DR} is the flexural tensile strength of the specimen after n freeze–thaw cycles (MPa) and Q_C is the initial flexural tensile strength of the specimen before freeze–thaw cycles (MPa).

(4) Fatigue test

The fatigue test was carried out according to T0856-2009 in JTG E51-2009 [20]. The fatigue life was used to characterize the fatigue performance of the material. The fatigue equation was calculated by the regression of Equation (10).

$$\lg N = a + b\sigma/S \quad (10)$$

where N is the number of load actions, σ is the load (N), σ/S is the strength-stress ratio, S is the flexural tensile strength of the specimen (MPa), and a and b are regression coefficients.

(5) Residual toughness evaluation

The standard toughness evaluation method was carried out for ASTM C1018-97 [21]. The toughness indices I_5, I_{10}, and I_{20} were used to evaluate the material toughness by the degree of deviation between the material and the ideal elastic–plastic body. The residual strength indices $R_{5,10}$ and $R_{10,20}$ were used to evaluate their plastic properties. Based on the measured load and deflection data, the residual toughness of PVA fiber cement-stabilized macadam was calculated according to the standard toughness evaluation method. The calculation steps are as follows.

① The area S_1, S_2, S_3, and S_4 enclosed by the corresponding load–deflection curve of the specimen at the corresponding deflection is calculated, where S_1, S_2, S_3, and S_4 are the areas enclosed by the first crack deflection δ to the origin, δ to 3δ, 3δ to 5.5δ and 5.5δ to 10.5δ regarding the load–deflection curves and the horizontal coordinate axis, respectively.

② The toughness indices I_5, I_{10} and I_{20} are calculated according to Equations (11)–(15), respectively.

$$I_5 = (S_1 + S_2)/S_1 \quad (11)$$

$$I_{10} = (S_1 + S_2 + S_3)/S_1 \quad (12)$$

$$I_{20} = (S_1 + S_2 + S_3 + S_4)/S_1 \quad (13)$$

③ The residual strength indices $R_{5,10}$ and $R_{10,20}$ are calculated according to Equations (14) and (15).

$$R_{5,10} = 20 \times (I_{10} - I_5) \quad (14)$$

$$R_{10,20} = 10 \times (I_{20} - I_{10}) \quad (15)$$

3. Results and Discussion

3.1. Shrinkage Characteristics of PVA Fiber Cement-Stabilized Macadam

3.1.1. Dry Shrinkage Performance

The dry shrinkage test of cement-stabilized macadam was used to compare and analyze the effect of incorporating PVA fiber on the dry shrinkage performance of cement-stabilized macadam, as shown in Figure 2.

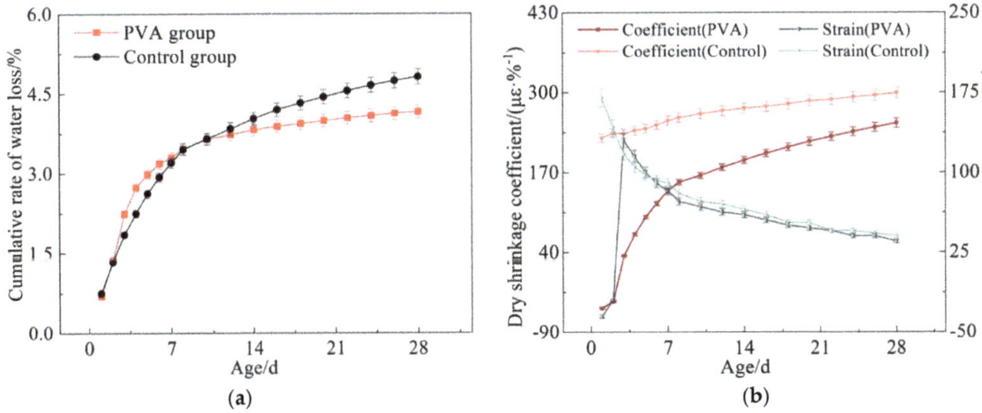

Figure 2. Changes in the dry shrinkage properties of cement–stabilized macadam: (**a**) variation pattern of water loss rate with age; (**b**) coefficient/strain of dry shrinkage variation law with age.

In Figure 2a, both PVA fiber cement-stabilized macadam and ordinary cement-stabilized macadam show a trend of an increasing rate of water loss with age, then gradually slow down and stabilize. The rate of water loss in the first 10 days of the age of the mixture increased faster, among which the rate of water loss of PVA fiber cement-stabilized macadam was relatively faster. The cumulative rate of water loss of PVA fiber cement-stabilized macadam at the age of 28 days was 13.7% lower compared with ordinary cement-stabilized macadam.

In Figure 2b, the dry shrinkage coefficient of PVA fiber cement-stabilized macadam changes with age in the same way as the rate of water loss. The dry shrinkage strain increases rapidly from negative values in the first 3 days, and then gradually decreases and stabilizes. The dry shrinkage coefficient of PVA fiber cement-stabilized macadam is reduced by 16.4% compared with ordinary cement-stabilized macadam at the age of 28 days, which indicates that the incorporation of PVA significantly improves the dry shrinkage performance of cement-stabilized macadam. This is probably because the water originally adsorbed on the surface of the minerals and binders in the cement-stabilized macadam is partially adsorbed on the surface of the PVA fiber after the fiber is incorporated. The water on the surface of the fiber continued to dissipate with age. In turn, the rate of water loss and dry shrinkage strain increased rapidly in the early stage, and gradually leveled off with age.

3.1.2. Temperature Shrinkage Performance

The temperature shrinkage test of cement-stabilized macadam was used to compare and analyze the effect of incorporating PVA fiber on the temperature shrinkage performance of cement-stabilized macadam, with the help of the temperature shrinkage strain and temperature shrinkage coefficient, as shown in Figure 3.

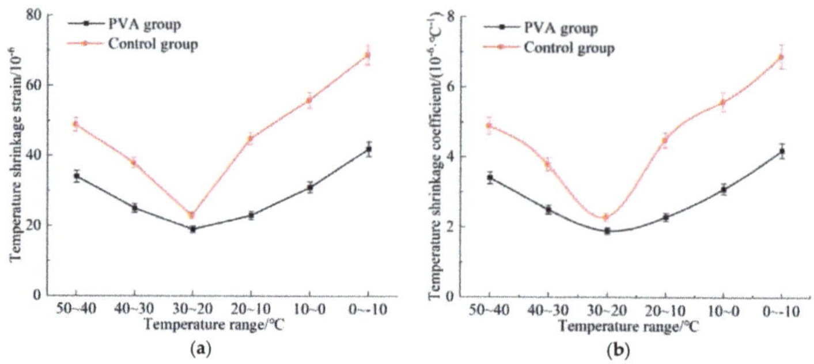

Figure 3. Changes in the temperature shrinkage performance of cement–stabilized macadam: (**a**) temperature shrinkage strain at different temperatures; (**b**) temperature shrinkage coefficient at different temperatures.

In Figure 3, the temperature shrinkage strain and temperature shrinkage coefficient of both types of cement-stabilized macadam decreased with decreasing temperature, when the temperature was higher than 20 °C. The temperature shrinkage strain and temperature shrinkage coefficient of the mixes increased significantly with decreasing temperature below 20 °C. The strain and coefficient of temperature shrinkage of ordinary cement-stabilized macadam are greater than PVA cement-stabilized macadam, and the rate of change of the curve is relatively large. The strain and coefficient of temperature shrinkage of the mix after incorporating PVA fiber decreased by 17.4~48.9% in the temperature range of −10~50 °C, among which the temperature shrinkage strain and temperature shrinkage coefficient of cement-stabilized macadam reached the minimum in the temperature range of 30~20 °C. A decrease of 17.4% was recorded after the incorporation of PVA fibers. The results showed that the low-temperature sensitivity of PVA fiber improved the temperature shrinkage of cement-stabilized macadam, due to the high- and low-temperature changes. In addition, the PVA fiber exerted a certain reinforcing effect and inhibited the early cracking of cement-stabilized macadam due to temperature shrinkage.

3.2. *Low-Temperature Residual Toughness of PVA Fiber Cement-Stabilized Macadam*

3.2.1. Low-Temperature Flexural Toughness

Cement-stabilized macadam should have good low-temperature toughness to withstand long-term environmental temperature changes, wet and dry conditions, freeze–thaw cycles, etc. [22–24]. Therefore, the freeze–thaw cycle and three-point bending test were used to compare and evaluate the changes in flexural tensile strength, bearing capacity, and deflection of PVA fiber cement-stabilized macadam before and after the freeze–thaw cycle, as shown in Figure 4, and the fracture patterns of the specimens are shown in Figure 5.

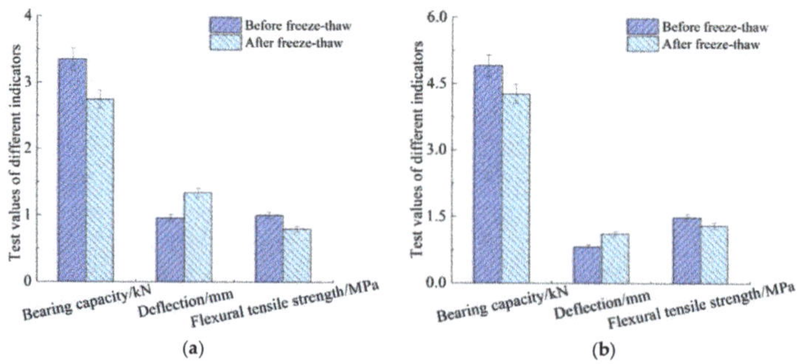

Figure 4. Comparison of the toughness of different types of cement–stabilized macadam: (**a**) control group; (**b**) PVA group.

Figure 5. Specimen fracture pattern: (**a**) ordinary cement−stabilized macadam; (**b**) PVA fiber cement−stabilized macadam.

From Figure 4, it can be observed that the bending toughness of cement-stabilized macadam significantly improved after the incorporation of PVA fiber. The flexural tensile strength without freeze–thaw and after freeze–thaw was increased by 47.9%, and 63.1%, respectively, and the residual flexural strength was increased by 10.3%. The bearing capacity of PVA fiber cement-stabilized macadam before freeze–thaw was increased by 46.6% compared with unadulterated fiber specimens and the bearing capacity decreased by 13.6% after freeze–thaw and increased by 55.3% compared with unadulterated fiber specimens. The deflection was reduced by 14.6% compared with unadulterated fiber specimens and decreased by 17.2% compared with unadulterated fiber specimens after freeze–thaw.

From Figure 5, according to the characteristics of the fracture surface, it can be found that the fracture surface of ordinary cement-stabilized macadam is close to a straight line, which shows the damage behavior of the aggregate and cement bonding interface. On the other hand, the fracture surface of PVA fiber cement-stabilized macadam is an irregular folded line, which indicates that fracture damage hysteresis has occurred during the fracture, and the fracture interface of the PVA fiber cement-stabilized macadam specimens displays pull-out fibers, which indicates that the fiber prolongs the critical state process of cement-stabilized macadam fractures, and the fiber force slows down the cracking rate of the specimens, meaning that the energy required for specimen cracking is increased. The frictional resistance between the PVA fibers and cement matrix plays a key role in resisting the rapid destruction of the load, which indicates that PVA fibers can effectively improve the low-temperature toughness of cement-stabilized macadam.

3.2.2. Residual Toughness Evaluation

To further quantify the effect of PVA fiber on the low-temperature toughness of cement-stabilized macadam, the low-temperature toughness of PVA fiber cement-stabilized macadam was systematically evaluated using the toughness indices I_5, I_{10}, I_{20}, and residual strength indices $R_{5,10}$ and $R_{10,20}$ with the help of the ASTM C1018-97 standard toughness evaluation method, as shown in Figures 6 and 7.

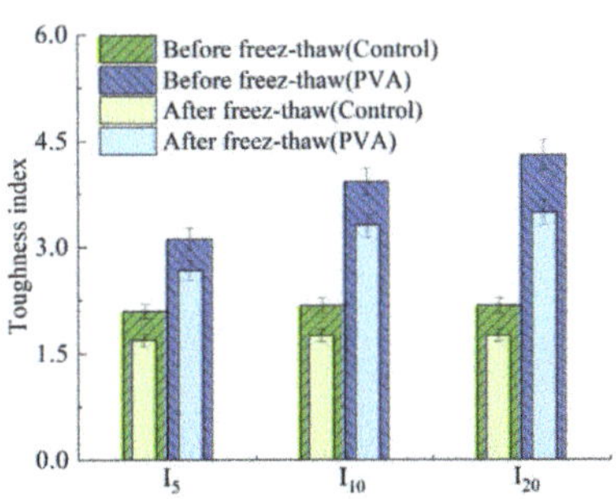

Figure 6. Toughness index.

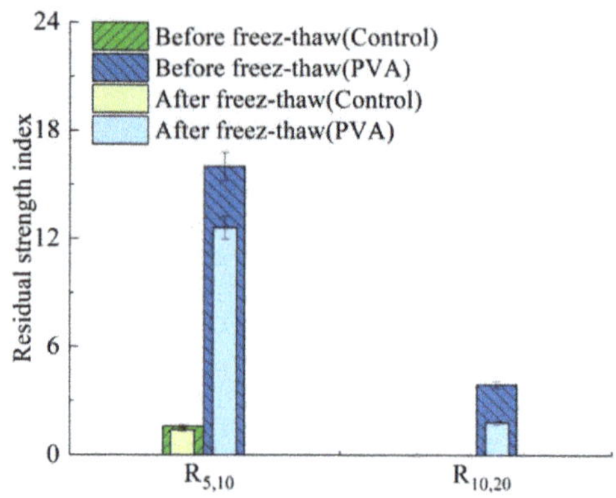

Figure 7. Residual strength index.

From Figures 6 and 7, it can be observed that the toughness index and residual strength index of cement-stabilized macadam increased significantly after mixing with PVA fiber. The toughness indices I_5, I_{10}, and I_{20} increased by 48.6%, 79.8%, and 97.7%, respectively, and the residual strength indices $R_{5,10}$ and $R_{10,20}$ increased by more than 100%. The toughness evaluation index of cement-stabilized macadam decreased to some extent after the freeze–thaw cycles. Among them, the toughness indices of PVA fiber cement-stabilized macadam I_5, I_{10}, and I_{20} decreased by 14.1%, 15.6%, and 19%, respectively, and the residual strength indices $R_{5,10}$ and $R_{10,20}$ decreased by 21.3% and 53.8%, respectively. The residual toughness indices increased by 58.6%, 88.1%, and 98.3%, respectively, compared with ordinary cement-stabilized macadam after freeze–thaw, and the residual strength index increased by more than 100%. This indicates that the strong decrease in the mixture is slowed down by the crack-arresting effect of the flexible fibers after the appearance of cracks, while more energy needs to be dissipated for further expansion of the cracks, thus effectively improving the flexural toughness of the cement-stabilized macadam.

3.3. *Fatigue Performance Analysis of PVA Fiber Cement-Stabilized Macadam Based on Weibull Distribution*

Cement-stabilized macadam needs to have sufficient fatigue resistance to withstand vehicle loading for a long period [25]. The indoor fatigue tests showed that the fatigue life of cement-stabilized macadam varied at the same stress level, and the test results were discrete; therefore, it was difficult to analyze the fatigue life at different stress levels. In recent years, mathematical analysis methods have been widely used in the field of engineering [26]. Therefore, to accurately and objectively compare the effects of PVA fibers on the fatigue life of cement-stabilized macadam at different stress levels, the fatigue life of the specimens of cement-stabilized macadam at 28 days was tested by using the linear regression method combined with Weibull distribution [27]. The reliability and the prediction equations of the fatigue life of cement-stabilized macadam were calculated according to Equations (16) and (17).

$$P = 1 - \frac{i}{k+1} \tag{16}$$

$$N = N_a \times [ln(\frac{1}{P})]^{\frac{1}{b}} \tag{17}$$

where *i* is the *i*th specimen number, and *k* is the number of specimens with the same stress level.

The fatigue life of cement-stabilized macadam under different stress intensity ratios was calculated by linear regression, as shown in Figure 8.

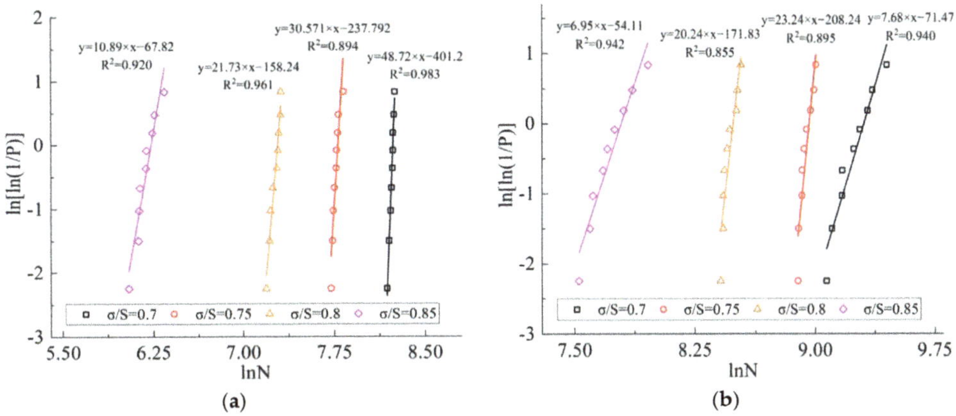

Figure 8. Fatigue life Weibull distribution fitting comparison: (**a**) ordinary cement−stabilized macadam; (**b**) PVA fiber cement−stabilized macadam.

In Figure 8, the average fatigue life of PVA fiber cement-stabilized macadam improved between 178% and 368% compared with non-fiber cement-stabilized macadam for the stress intensity ratios of 0.7, 0.75, 0.8, and 0.85. Under the same stress level, the slope and intercept of the fitted equation of the fatigue life of cement-stabilized macadam with PVA fiber are relatively small. The fatigue life is less sensitive to the change in stress level, and the actual fatigue life of cement-stabilized macadam under four stress levels has a significant linear relationship with the fatigue life of Weibull distribution when the reliability is P, which shows that the fatigue life of cement-stabilized macadam obeys the two-parameter Weibull distribution. Therefore, the regression coefficients were substituted into Equation (16) to calculate the fatigue life versus reliability curves for different types of cement-stabilized macadam, which are shown in Figure 9.

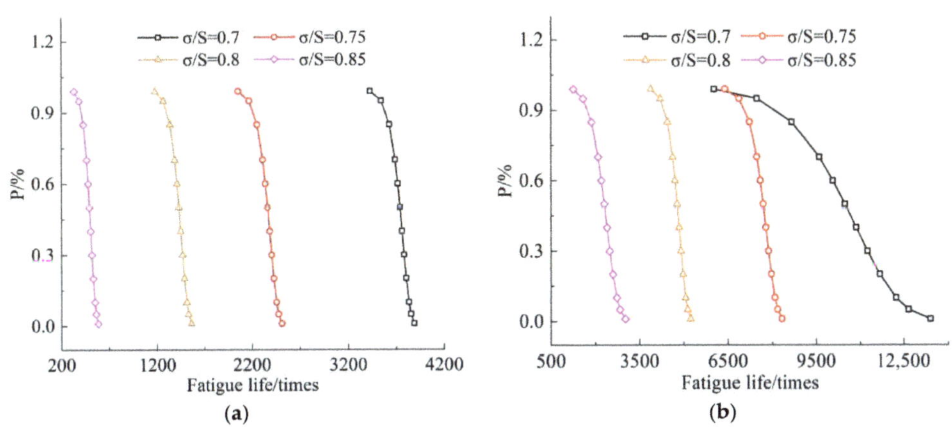

Figure 9. Reliability–fatigue life variation law: (**a**) ordinary cement−stabilized macadam; (**b**) PVA fiber cement−stabilized macadam.

In Figure 9, the fatigue life reliability of different cement-stabilized macadam under the same stress ratio gradually decreases with the increase in the number of load cycles. The fatigue life of PVA fiber cement-stabilized macadam is greater than ordinary cement-stabilized macadam at the same reliability. Compared with the measured fatigue life, the prediction error ranges from 0.3% to 4.9% at the stress ratio of 0.7, from 0.03% to 1.6% at the stress ratio of 0.75, from 0.3% to 2.2% at the stress ratio of 0.8, and from 0.12% to 3.5% at the stress ratio of 0.85. Therefore, suitable reliability should be determined for the fatigue-life analysis of cement-stabilized macadam in conjunction with the actual conditions.

Taking the fatigue life of cement-stabilized macadam at 50% and 95% reliability as an example, according to the fatigue equation recommended by T0856-1 from the JTG E51-2009 [20], the aforementioned Equation (10) can be established. Meanwhile, based on the Weibull distribution, the prediction equation for the fatigue life of cement-stabilized macadam is established as shown in Table 2, and the fitting results are shown in Figure 10.

Table 2. Prediction equations for fatigue life of different cement-stabilized macadam mixes.

Material Type	Reliability/%	Fatigue Life Equation	R^2
Control group	0.5	$\lg N = 17.594 - 13.201\sigma/S$	0.953
	0.95	$\lg N = 18.386 - 14.384\sigma/S$	0.949
PVA	0.5	$\lg N = 16.393 - 10.053\sigma/S$	0.966
	0.95	$\lg N = 16.348 - 10.302\sigma/S$	0.869

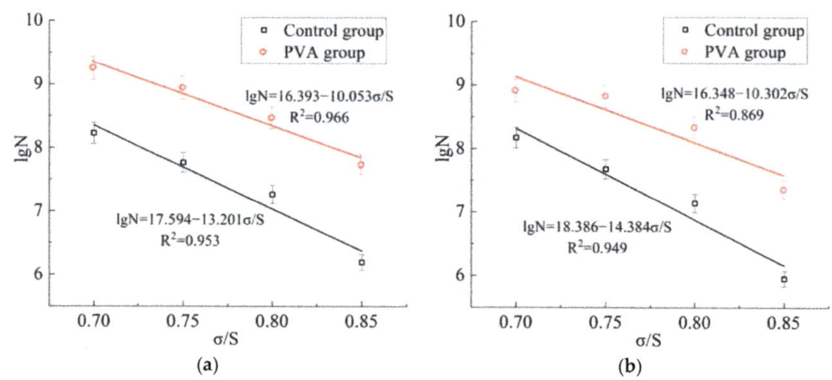

Figure 10. Fatigue life fitting curve with different reliability indices: (**a**) 50% reliability; (**b**) 95% reliability.

In Figure 10, except for the prediction coefficient of the fatigue life of PVA fiber cement-stabilized macadam with 95% reliability, the coefficient of determination of all the equations is greater than 0.94, and the fatigue equation of cement-stabilized macadam has a strong correlation with the test results. The fatigue life of PVA fiber cement-stabilized macadam under different reliabilities is higher than ordinary cement-stabilized macadam, which indicates that PVA fiber can significantly improve the fatigue resistance of cement-stabilized macadam. In addition, the fatigue life of PVA fiber cement-stabilized macadam under different reliability levels has some differences, for instance, when the reliability level is reduced from 95% to 50%, the fatigue life increases by 1.3~5.1% under different stress ratios. Accordingly, the appropriate reliability should be selected to accurately analyze the fatigue life of cement-stabilized macadam by combining the material type, road grade, and stress state in practical applications.

4. Conclusions and Prospects

(1) PVA fiber effectively improved the deformation characteristics and flexural toughness of cement-stabilized macadam. The cumulative water loss rate and dry shrinkage coefficient of PVA fiber cement-stabilized macadam decreased by 13.7% and 16.4%, respectively, when compared with those without fiber at 28 days of age. The strain and coefficient of temperature shrinkage of PVA fiber cement-stabilized macadam decreased by 17.4~48.9%, compared with those without fiber in the temperature range of −10~50 °C. The residual flexural tensile strength and low-temperature bearing

capacity increased by 10.3% and 55.3%, respectively, after fiber incorporation. The deflection after the freeze–thaw cycle decreased by 17.2%, the residual toughness indices increased by 58.6%, 88.1%, and 98.3% and the residual strength index increased by more than 100%.

(2) PVA fiber improved the fatigue performance of cement-stabilized macadam and its fatigue life obeyed the two-parameter Weibull distribution. Its fatigue life under different reliabilities was also higher than the specimens without fiber. The establishment of the prediction equation for the fatigue life of the mix under different reliabilities can reflect the improved effect of PVA fiber on the fatigue performance of cement-stabilized macadam more accurately.

(3) In this study, the durability performance of PVA fiber cement-stabilized macadam was analyzed mainly based on indoor tests, and the flexural toughness and fatigue performance were evaluated by combining standard toughness evaluation methods and mathematical statistics, but the durability enhancement mechanism has not yet been investigated. In the future, the durability enhancement mechanism of PVA fiber cement-stabilized macadam must be investigated in depth, and a test road will be needed for long-term performance monitoring.

Author Contributions: Conceptualization, S.T. and C.W.; methodology, Q.Z.; validation, S.T.; formal analysis, F.C.; investigation, Y.H.; resources, C.W.; data curation, Q.Z.; writing—original draft preparation, S.T.; writing—review and editing, C.W.; supervision, F.C.; project administration, C.W. All authors have read and agreed to the published version of the manuscript.

Funding: This research was sponsored by the Innovation Capability Support Program of Shaanxi (No.2022TD-07).

Institutional Review Board Statement: Not applicable.

Informed Consent Statement: Not applicable.

Data Availability Statement: All data, models, and codes generated or used during the study appear in the published article.

Conflicts of Interest: The authors declare no conflict of interest.

References

1. Wen, P.H.; Wang, C.H.; Song, L.; Niu, L.; Chen, H. Durability and Sustainability of Cement-Stabilized Materials Based on Utilization of Waste Materials: A Literature Review. *Sustainability* **2021**, *13*, 11610. [CrossRef]
2. Wang, C.; Li, Y.; Wen, P.; Zeng, W.; Wang, X. A comprehensive review on mechanical properties of green controlled low strength materials. *Constr. Build. Mater.* **2023**, *363*, 129611. [CrossRef]
3. Liu, L.Q.; Wang, C.H.; Liang, Q. Preparation of a Heat Insulation Bonding Layer for Roads and its Heat Insulation Effect. *J. Clean. Prod.* **2022**, *365*, 132828. [CrossRef]
4. Farhan, A.H.; Dawson, A.R.; Thom, N.H. Recycled Hybrid Fiber-Reinforced & Cement-Stabilized Pavement Mixtures: Tensile Properties and Cracking Characterization. *Constr. Build. Mater.* **2018**, *179*, 488–499.
5. Wei, J.; Wang, Y.; Li, X.; Jia, Z.; Qiao, S.; Zhang, Q.; Du, J. Effect of porosity and crack on the thermoelectric properties of expanded graphite/carbon fiber reinforced cement-based composites. *Int. J. Energy Res.* **2020**, *44*, 6885–6893. [CrossRef]
6. Chen, Q.; Wang, C.H.; Li, Y.W.; Feng, L.; Huang, S. Performance Development of Polyurethane Elastomer Composites in Different Construction and Curing Environments. *Constr. Build. Mater.* **2023**, *365*, 130047. [CrossRef]
7. Huang, Y.J. Composition Design and Performance Study of PVA Fibre Cement Stabilized Gravel Base. Master's Thesis, Chang'an University, Xi'an, China, 2022.
8. Alam, B.; Yaman, I.O. Fatigue Performance of PVA Fibre Reinforced Cementitious Composite Overlays. *Int. J. Pavement Eng.* **2019**, *22*, 822–828. [CrossRef]
9. Zhang, P.; Yuan, P.; Guan, J.F.; Guo, J. Fracture Behavior of Multi-Scale Nano-SiO_2 and Polyvinyl Alcohol Fiber Reinforced Cementitious Composites under the Complex Environments. *Theor. Appl. Fract. Mech.* **2022**, *122*, 103584. [CrossRef]
10. Kim, K.W.; Yu, C.; Han, J.W.; Park, C.G. Strength and Durability of Rapid Set PVA Fiber Reinforced LMC for Pavement Repair. *Polym. Polym. Compos.* **2019**, *27*, 179–188. [CrossRef]
11. Huang, Z.Q.; Xu, G.T.; Liu, X. Fracture Performance of PVA Fiber Cementitious Composites under Freeze-Thaw Cycles. *J. Shenyang Univ. Technol.* **2018**, *40*, 352–356.

12. Zhang, P.; Wang, L.; Zheng, Y.X. Study on Flexural Toughness Properties of PVA Fiber Reinforced Nano-SiO_2 Cement Composites. *New Build. Mater.* **2021**, *48*, 81–84.
13. Zhang, P.; Kang, Y.L.; Guo, J.J. Fracture Properties of Nano-SiO_2 and PVA Fiber Reinforced Cementitious Composites. *J. Build. Mater.* **2021**, *24*, 908–915.
14. Liu, T.Y.; Zhang, P.; Li, Q.F.; Hu, S.W.; Ling, Y.F. Durability Assessment of PVA Fiber-Reinforced Cementitious Composite Containing Nano-SiO_2 Using Adaptive Neuro-Fuzzy Inference System. *Crystals* **2020**, *10*, 347. [CrossRef]
15. Li, Y.J. Study on Crack Resistance and Durability of Cement Stabilized Macadam Mixture with PVA Fiber. *Highw. Eng.* **2020**, *45*, 180–188+219.
16. Yu, J.Y.; Liang, N.X.; Tong, P.; Tang, R.X. Fatigue Life Analysis of Polyvinyl Alcohol Fiber Modified Cement Stability Macadam Base Pavement. *Bull. Chin. Ceram. Soc.* **2019**, *38*, 2408–2413+2419.
17. Zhao, Y.; Yang, X.; Zhang, Q.Y.; Liang, Y.; Xiang, Y.; Qin, M. Crack Resistance and Mechanical Properties of Polyvinyl Alcohol Fiber-Reinforced Cement-Stabilized Macadam Base. *Adv. Civ. Eng.* **2020**, *2020*, 6564076. [CrossRef]
18. Wang, X.D. *JTG/T F20-2015*; Technical Guidelines for Construction of Highway Roadbases. Ministry of Transport of the People's Republic of China: Beijing, China, 2015.
19. Han, L.; Cai, H.F.; Chen, X.; Zheng, C.; Guo, W. Study of UHMWPE Fiber Surface Modification and the Properties of UHMWPE/Epoxy Composite. *Polymers* **2020**, *12*, 521. [CrossRef]
20. Wang, X.D.; Li, M.J.; Sha, A.M.; Wang, S.Y.; Zhou, X.Y.; Shen, G.H.; Lu, K.J. *JTG E51-2009;* Test Methods of Material Stabilized with Inorganic Binders for Highway Engineering. Research Institute of Highway Ministry of Transport: Beijing, China, 2009.
21. *ASTM C1018-97*; Standard Test Method for Flexural Toughness and First-Crack Strength of Fiber Reinforced Concrete (Using Beam with Third-Point Loading). ASTM: Philly, PA, USA, 1997.
22. Chen, Q.; Wang, C.H.; Yu, S.X.; Song, Z.; Fu, H.; An, T. Low-Temperature Mechanical Properties of Polyurethane-Modified Waterborne Epoxy Resin for Pavement Coating. *Int. J. Pavement Eng.* **2022**. [CrossRef]
23. Gu, D.; Liu, H.; Gao, X.; Huang, D.; Zhang, W. Influence of Cyclic Wetting-Drying on the Shear Strength of Limestone with a Soft Interlayer. *Rock Mech. Rock Eng.* **2021**, *54*, 4369–4378. [CrossRef]
24. Wang, L.Q.; Yin, Y.P.; Huang, B.L.; Dai, Z. Damage Evolution and Stability Analysis of the Jianchuandong Dangerous Rock Mass in the Three Gorges Reservoir Area. *Eng. Geol.* **2019**, *265*, 105439. [CrossRef]
25. Wang, S.; Wang, C.H.; Yuan, H.Z.; Ji, X.; Yu, G.; Jia, X. Size Effect of Piezoelectric Energy Harvester for Road with High Efficiency Electrical Properties. *Appl. Energy* **2023**, *330*, 120379. [CrossRef]
26. Wang, Z.Y.; Thomas, B.; Zhang, W.G.; Gu, D. A Novel Random Angular Bend (RAB) Algorithm and DEM Modeling of Thermal Cracking Responses of Sandstone. *Geomech. Energy Environ.* **2022**, *32*, 100335. [CrossRef]
27. Jiang, Y.J.; Yuan, K.J.; Deng, C.Q.; Tian, T. Fatigue Performance of Cement-Stabilized Crushed Gravel Produced Using Vertical Vibration Compaction Method. *J. Mater. Civ. Eng.* **2020**, *32*, 04020318. [CrossRef]

sustainability

Review

Materials and Performance of Asphalt-Based Waterproof Bonding Layers for Cement Concrete Bridge Decks: A Systematic Review

Feng Gao [1], Xuan Gao [2], Qian Chen [3,*], Yanduo Li [3], Zhiwei Gao [4] and Chaohui Wang [3,*]

1 Sino Nuclear Northwest Construction Group Co., Ltd., Xi'an 710054, China
2 CCCC First Highway Consultants Co., Ltd., Xi'an 710075, China
3 School of Highway, Chang'an University, Xi'an 710064, China
4 School of Information Engineering, Xizang Minzu University, Xianyang 712082, China
* Correspondence: 2016121160@chd.edu.cn (Q.C.); wchh0205@chd.edu.cn (C.W.)

Abstract: As an important part of the bridge deck pavement system, the waterproof bonding layer plays a vital role in ensuring the integrity and durability of the pavement structure. Asphalt-based waterproof bonding materials have attracted extensive attention from researchers due to their low cost and good combination with asphalt surfaces. However, the existing research results of asphalt waterproof bonding layers are confused and there is a lack of systematic summaries. In addition, there are significant differences in the type, specification, performance, evaluation method, and evaluation index of asphalt materials. The performance evaluation indexes and methods of asphalt waterproof bonding materials need to be further studied and improved. To further promote the research and development of asphalt waterproof bonding layer materials, in this paper, the relevant specifications for the waterproof bonding layer of roads and bridges in China were systematically combed, the key performance index requirements in different specifications were compared and evaluated, the research trends of the asphalt waterproof bonding layer in China and its application in engineering construction were comprehensively reviewed, the performance of different asphalt waterproof bonding materials were systematically investigated, and the construction technology and economy of different asphalt waterproof bonding layer materials were analyzed. This paper provides a useful reference for the specification improvement and quality control of asphalt waterproof bonding layer.

Keywords: road materials; bridge deck pavement; asphalt-based waterproof bonding layer; research progress; working properties

check for updates

Citation: Gao, F.; Gao, X.; Chen, Q.; Li, Y.; Gao, Z.; Wang, C. Materials and Performance of Asphalt-Based Waterproof Bonding Layers for Cement Concrete Bridge Decks: A Systematic Review. *Sustainability* **2022**, *14*, 15500. https://doi.org/10.3390/su142315500

Academic Editor: Rui Micaelo

Received: 1 September 2022
Accepted: 15 November 2022
Published: 22 November 2022

1. Introduction

A waterproof bonding layer acts as a functional layer between the bridge deck and the paving layer, acting as a bonding and waterproofing layer. A good waterproof bonding layer not only ensures the integrity of the paving structure, but also plays a part in stress absorption, inhibits surface cracking, and offers other effects [1,2]. At present, the commonly used bridge deck waterproof bonding layer materials in China are mainly categorized as asphalt and reactive resin types, and asphalt materials, because of their own water repellent, low cost, and good adhesion with the surface layer, are widely used in the field of bridge deck pavements, while the development of modified asphalt, so that it not only retains the original advantages of asphalt materials, but also has a number of significant characteristics of modified asphalt, is ongoing. In recent years, researchers have carried out a series of studies on the performance of different modified asphalt-based waterproof bonding materials. Scholars represented by Haynes proposed a variety of modified asphalt-based waterproof adhesive materials, with self-healing, crack resistance, impermeability, and other aspects to enhance the performance of waterproof materials [3–6]. Scholars represented by Oh evaluated the service performance of different types of waterproof adhesive layer [7–9]. Scholars represented by Hailesilassie analyzed the bonding and debonding mechanism of waterproof adhesive materials in different situations based on laboratory tests [10–14]. In summary, it can be seen that the current research is mainly focused on the development of waterproofing binder materials and interlayer bonding performance, but the results are complicated, scattered, lack systematic sorting, and engineering applications have not formed a complete system. In order to further promote the research and application of asphalt-based waterproof adhesive layer, it is necessary to more systematically comb the changing rules of the working performance of different asphalt-based waterproof adhesive layers, so as to conduct further targeted research.

Based on above reasons, in this paper, the relevant specifications of asphalt waterproof bonding layer for roads and bridges in China were systematically sorted out, the key technical indicators of waterproof bonding layer in different specifications were compared and evaluated, the research trends and the engineering applications of bridge deck waterproof bonding layer in China were comprehensively investigated, the performance of different asphalt waterproof bonding layer materials was comprehensively summarized, and the construction technology and economy of different asphalt waterproof bonding layer materials were compared and analyzed. This work provides a reference for revising and improving the relevant specifications of asphalt waterproof bonding materials, and lays a foundation for subsequent engineering applications and targeted in-depth research.

2. Survey and Evaluation of Asphalt-Based Waterproof Bonding Layer Specifications

The asphalt-based waterproof bonding layer is effectively bonded with the bridge deck through physical action. The material softens or melts with the increase in temperature, and solidifies with the decrease in temperature. In the project, the asphalt-based waterproof bonding layer is divided into coiled material, coating film, and crushed stone seal according to the structure type [15–17]. Because of the different structures of bridge deck pavement combinations and the wide variety of waterproof bonding materials and specifications, there are significant differences in their corresponding performance index requirements and evaluation methods. In order to comprehensively and objectively evaluate the technical performance of the waterproof bonding layer, this paper comprehensively investigated the relevant specifications of waterproof bonding layer materials for roads and bridges in China, and compared their key performance indicators and specific requirements, as shown in Table 1.

Table 1. Relevant specifications of bonding layers for roads and bridges.

Serial Number	Specification Name	Specification Number
1	Atactic polypropylene(APP) asphalt waterproof roll for roads and bridges	JT/T 536-2004
2	Waterproofing coatings for concrete bridges and road surfaces	JC/T 975-2005
3	Emulsified asphalt waterproof coating	JC/T 408-2005
4	Modified bituminous waterproofing sheets for concrete bridge decks and other concrete surfaces trafficable by vehicles	JC/T 974-2005
5	Technical guide to the design and construction of highway steel box girder bridge decks	[2006] NO.274
6	Standard specification for waterborne epoxy-binder waterproof coating for concrete bridges	DB32/T 2285-2012
7	Technical specifications for asphalt pavement interlayer treatment of highways	DB61/Z 917-2014
8	General specifications of epoxy asphalt materials for paving roads and bridges	GB/T 30598-2014
9	Water quality asphalt waterproof coating for highways	JT/T 535-2015
10	Technical specifications for the construction of ERS steel deck pavements	DB33/T 2012-2016
11	Specifications for the design and construction of pavements on highway steel deck bridges	JTG/T3364-02-2019
12	Technical specification for rapid construction of a waterproof-bonding layer on concrete bridge decks	DB11/T 1680-2019

According to the specifications in Table 1, at the beginning of the application of a waterproof bonding layer for bridge deck pavements, building waterproof coiled materials, the asphalt film, and other materials were mostly used, but the effect was not ideal. At present, the composite waterproof system is mainly composed of a synchronous chip seal, high-performance epoxy materials, and a variety of materials. Considering the characteristics of bonding layer materials and the influence of construction factors, it is proposed that the key technical indicators of a waterproof bonding layer should include drying time, heat resistance, low temperature flexibility, impermeability, and bonding performance.

2.1. Drying Time

There are five specifications in the survey that limit the drying time of asphalt waterproof bonding materials, mainly for solvent asphalt and water emulsion asphalt materials (see Figure 1 for details).

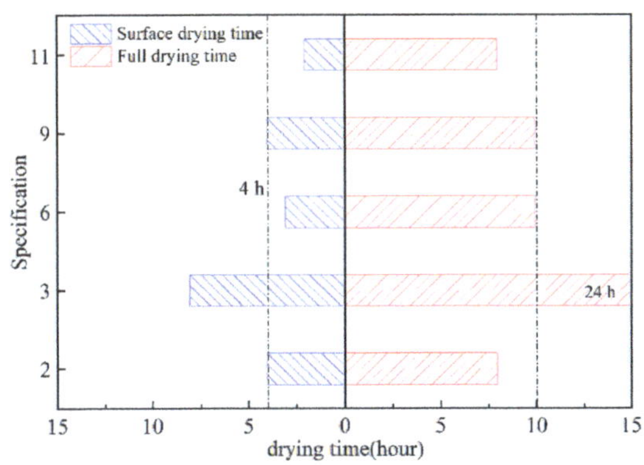

Figure 1. Comparison of relevant specifications for drying time of waterproof bonding layer materials.

Different specifications for asphalt-based waterproof bonding materials' drying time requirements are broader. Among them, specifications 2, 6, 9, and 11 have a surface drying time of basically less than 4 h and a hard drying time of basically less than 8 or 10 h. Compared with the above specifications, the drying time in specification 3 is the longest, and the surface drying time is consistent with the hard drying time in specification 2, both of which are 8 h, while the hard drying time is up to 24 h. As far as the different asphalt types were concerned, solvent asphalt bonding materials have the most stringent drying time requirements; specification 11 requires a surface drying time of less than 2 h, and specifications 2, 3, 6, and 9 require a surface drying time of 3 h, 4 h, or 8 h for emulsified asphalt bonding materials. This indicates that solvent-based bonding materials evaporate into film faster and are relatively more demanding than water-emulsion materials. Generally speaking, the surface drying time is short, the materials easily solidify quickly, and the immediate construction rarely produces the sagging phenomenon, meaning that the project requirements can be met in a short time. However, a too short surface drying time will affect the adhesion of the bonding layer, which will lead to a reduction in the bonding performance between layers. Therefore, the surface drying and hard drying time can be determined according to the actual situation to obtain a better interlayer bonding effect.

In summary, when solvent asphalt waterproof bonding materials are applied to bridge deck pavements, the surface drying time shall be less than 2 h and the hard drying time shall be less than 8 h, while the emulsion asphalt bonding materials shall meet the requirements of a surface drying time of less than 4 h and a hard drying time of less than 10 h. According to specification 2, a thermal-modified asphalt binder shall meet the requirements that the surface drying time is less than 4 h and the hard drying time is less than 8 h.

2.2. Heat-Resistant Properties

There are seven specifications in China that put forward requirements on the heat resistance of asphalt waterproof bonding layer materials, mainly for modified asphalt waterproof rolls, water emulsion asphalt materials, heat-modified asphalt materials, and epoxy asphalt. See Figure 2 for details.

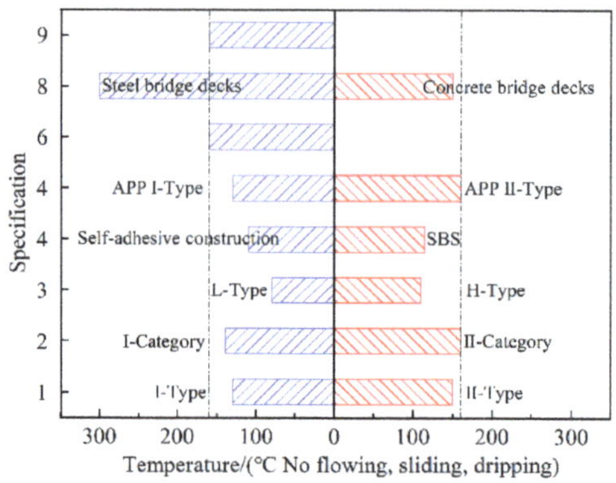

Figure 2. Comparison of relevant specifications for heat-resistant properties of waterproof bonding layer materials.

In Figure 2, the four specifications for heat-resistant property requirements are 160 °C with no flow, sliding, and dripping, while this is also in addition to specification 8 in the steel bridge deck epoxy asphalt waterproof bonding layer, namely that 300 °C does not melt outside the highest requirements. Parts of the specification requirements for heat-resistant properties of 150 °C and below, especially in the specification 3 requirements of high-performance (H-type) waterproof bonding material with a heat resistance of 110 °C, are relatively general for the visible water emulsion type asphalt waterproof bonding material, with a low performance (L-type) of only 80 °C. However, with the increase in engineering requirements and the development of modified asphalt, waterproof bonding material requirements for heat-resistant properties have also gradually increased. For example, rubber-modified asphalt mixes can be paved at temperatures of around 180 °C, although there is a significant temperature difference between the waterproof bonding layer and the asphalt mix, and the lower thermal limits make it difficult to meet the growing demand. Therefore, in order to avoid the asphalt paving process due to high temperature caused by excessive softness and flow of waterproof bonding material caused by hot aggregate puncture phenomenon, while taking into account the appropriate softening of bonding material or the slight flow that is conducive to improving the adhesion, waterproof bonding material heat-resisting properties should meet the following technical conditions: 160 °C conditions, bonding material without sliding, flowing, dripping.

2.3. Low-Temperature Flexibility

There are six specifications in the survey that limit the low temperature flexibility of asphalt waterproof bonding layer materials, mainly for modified asphalt waterproof coiled materials and water emulsion asphalt materials, as shown in Figure 3.

In Figure 3, the low-temperature flexibility of asphalt materials was more strictly controlled, with more attention paid to the effects of heat treatment, corrosion, and aging on them. Under standard conditions, the three specifications for low-temperature flexibility are the same, with no cracks or fractures at −25/−15 °C. Specification 1 for hot mix asphalt pavement APP asphalt waterproofing membrane and specification 3 in the H-type emulsified asphalt material requirements is relatively low, at only −10 and 0 °C without cracking or fracture. Most high-performance water-based asphalt materials and SBS-modified asphalt waterproofing roll-roofing are mainly between −25–−20 °C. In different corrosive and aging environments, the requirements for low-temperature performance were mostly relaxed by 5 to 10 °C. As the temperature drops, the brittleness of the asphalt material increases, and shrinkage can lead to cracks or fractures, so this property was often considered in cold areas and the requirements could be relaxed in hot and humid climates. In summary, in hot and humid climatic conditions, requirements can be relaxed to −15 °C without cracks and fractures; cold climatic conditions should meet −25 °C without cracks and fractures; corrosion and aging conditions can be relaxed by 5 °C.

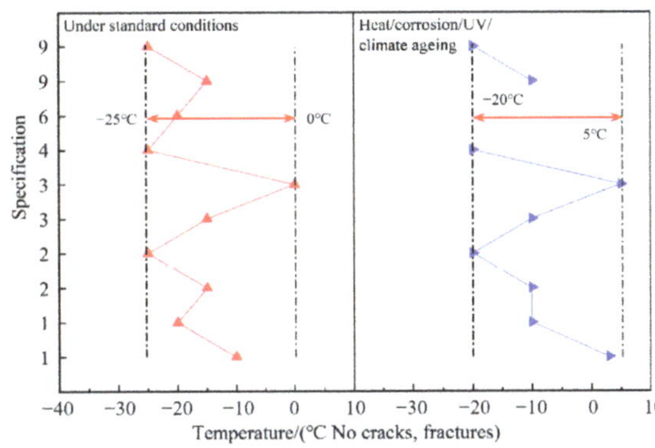

Figure 3. Comparison of relevant specifications for the low-temperature flexibility of waterproof bonding layer materials.

2.4. Waterproof Properties

A total of 10 specifications in China require waterproof properties for asphalt-based waterproof bonding layer materials, as shown in Figure 4.

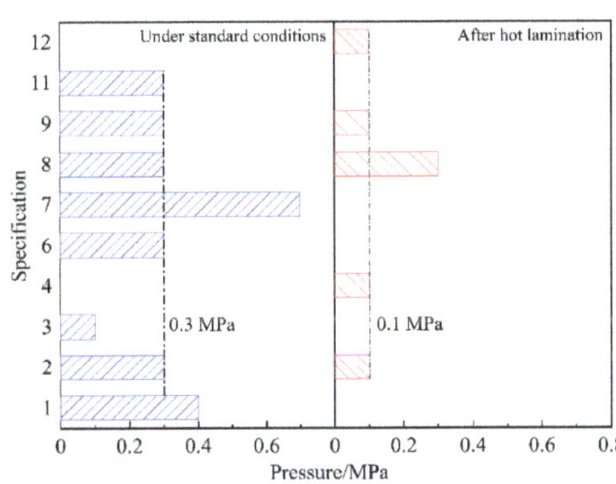

Figure 4. Comparison of relevant specifications for the waterproof properties of waterproof bonding layer materials.

In Figure 4, five specifications have the same waterproof performance requirements for asphalt materials, which is 0.3 MPa and that it is impermeable within 30 min. In other specifications, the same test conditions only require the film to maintain a high impermeable pressure; for example, the pressure in specification 1 and specification 7 is 0.4 and 0.7 MPa, respectively. This also shows that compared with the coating materials, such as emulsion asphalt, heat-modified asphalt, and solvent asphalt, the modified asphalt waterproof coiled material has higher requirements for water resistance and a better waterproofing effect. In addition, since the superstructure of the waterproof bonding layer is vulnerable to the impact of construction machinery or hot aggregate paving during the construction process, resulting in the asphalt film being punctured and damaged and affecting its performance, the five specifications have added hot rolling conditions to the waterproof performance test. Four of them require that the waterproof performance after hot rolling is 0.1 MPa, and that it is impermeable within 30 min. The highest requirement in specification 8 is 0.3 MPa, and that it is impermeable within 30 min. Considering that the resistance to construction damage of the waterproof bonding layer is extremely important, the waterproof performance requirements should generally meet the following technical conditions: 0.3 MPa after hot rolling, and 30 min impermeable.

2.5. Bonding Properties

Bonding performance is the most important road performance index of the waterproof bonding layer, which is generally characterized by adhesion pull-out, interlayer pull-out and interlayer shear. There are seven specifications in China that put forward requirements for the bonding performance of asphalt waterproof bonding layer materials, as shown in Figure 5.

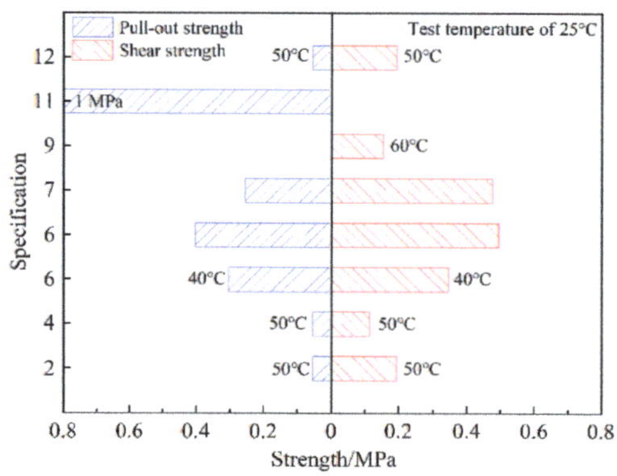

Figure 5. Comparison of relevant specifications for the bonding properties of waterproof bonding layers.

For the interlayer bonding properties of asphalt-based waterproof bonding materials, different specifications respectively limit the water emulsion asphalt, hot melt asphalt, modified asphalt roll, waterborne epoxy asphalt, SBS-modified asphalt seal, rubber-modified asphalt seal, solvent asphalt, and fiber reinforced rubber asphalt waterproof bonding layer. Analysis of Figure 5 shows that for the shear strength, specifications 6 and 7 require that waterborne epoxy asphalt, SBS-, and rubber-modified asphalt seals should be higher than 0.5 or 0.48 MPa at 25 °C. Specifications 2, 9, and 12 require high-temperature conditions (40–60 °C) for water emulsion asphalt and that the modified asphalt waterproof bonding layer should be higher than 0.16 or 0.2 MPa. Specification 4 for the modified asphalt membrane shear strength requirements requests minimum high-temperature conditions higher than 0.12 MPa. This is mainly due to the coil-type material on the concrete interface treatment requirements being high, and the coil material joints' fragile shear deformation resistance being poor, so its requirements are broader. For the pull strength, in addition to specification 11 in the solvent asphalt class of material requirements are relatively high, but the rest of the difference is not much, while specifications 6 and 7 require the waterproof bonding layer at 25 °C to have a pull strength higher than 0.25–0.4 MPa. Specifications 2, 4, and 11 require that the high temperature (50 °C) pull-out strength should be higher than 0.05 MPa.

In conclusion, when the modified asphalt is used as the waterproof bonding layer, pull strength should be higher than 0.3 MPa, and the shear strength should be higher than 0.05 MPa. When the emulsion asphalt is used as the waterproof bonding layer, the shear strength at 50 °C must be higher than 0.05 MPa. The bonding performance of waterborne epoxy asphalt, SBS-, and rubber-modified asphalt seal coat is excellent, so the requirements can be appropriately raised, that is, that the shear strength at 25 °C is not less than 0.5 MPa, and that the shear strength at high temperatures is not less than 0.35 MPa.

3. Investigation and Evaluation of the Working Properties of Asphalt-Based Waterproof Bonding Layer Materials

A comprehensive survey of China's roads and bridges with asphalt-based waterproof bonding layer material research dynamics and physical engineering application status, based on mathematical and statistical principles [18], was carried out. It summarized the application of asphalt-based waterproof bonding layer material technical indicators, clarified the level of performance of different asphalt-based waterproof bonding layer materials and their differences, and provided a reference for subsequent targeted in-depth research. For statistical purposes, data that were too fragmented were excluded from the statistics.

3.1. Basic Properties of Asphalt-Based Waterproof Bonding Layer Materials

According to the specifications, the basic properties of asphalt-based waterproof bonding materials mainly include pull-out strength, heat resistance, low temperature flexibility, water permeability resistance, and other properties. When applied to the surface of bridge concrete or steel plate, the amount of waterproof materials should be 0.7–1.0 kg/m^3 [19–32]. We have carried out investigation and statistics on the above properties of the waterproof bonding layer, as shown in Figure 6.

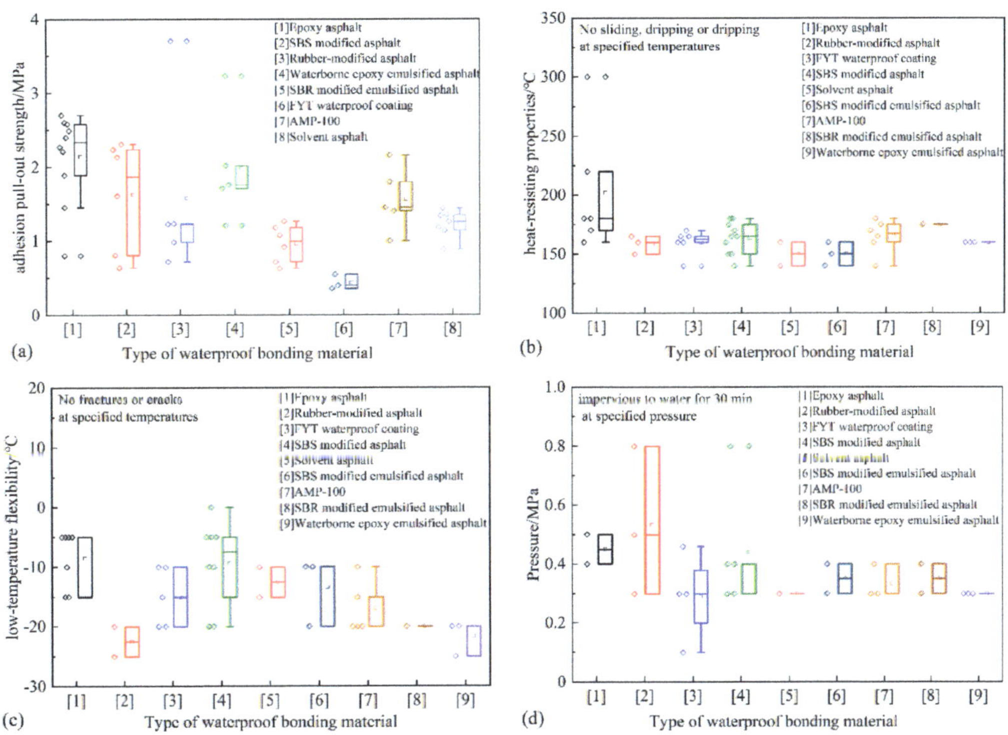

Figure 6. Basic properties of the asphalt waterproof bonding layer materials. (**a**) Mechanical properties, (**b**) heat-resistant properties, (**c**) low-temperature flexibility, and (**d**) waterproof properties.

In Figure 6a, the epoxy asphalt adhesion pull-out strength is the highest, with 75% of the epoxy asphalt data having a strength higher than 1.89 MPa, and mainly concentrated between 1.89~2.58 MPa. Waterborne epoxy emulsified asphalt adhesion pull-out strength has an average value of 1.99 MPa, the overall strength is slightly lower than the epoxy asphalt, and the data is mainly concentrated in the range of 1.71~2.02 MPa. Rubber-modified asphalt, SBS-modified asphalt, and AMP waterproof coating adhesion's pull-out strengths are relatively similar, in the range of 1.56~1.62 MPa. The SBR-modified emulsified asphalt and FYT waterproof coating's overall strength is low, the former having 50% of the data concentrated between 0.72~1.18 MPa, while the latter's values are mainly concentrated in 0.36~0.55 MPa. It can be seen that, in addition to waterborne epoxy emulsified asphalt, the mechanical properties of the remaining water-emulsified asphalt-like materials are low, which is mainly due to the small asphalt content as a result of the general emulsified asphalt-like materials' low viscosity, and the coating breaking because the emulsion curing film thickness is thin, which affects the adhesion with the concrete panel.

In Figure 6b, as a thermosetting material, epoxy asphalt has the best heat resistance, with its heat resistance mainly concentrated between 170 and 220 °C. The mean and median heat resistance of the remaining materials can meet the specification of 160 °C without flowing, sliding, or dripping. Analysis of Figure 6c shows that for low-temperature flexibility performance, the rubber asphalt temperature was mainly concentrated between −20 to −25 °C, which meets the requirements of specification 2 for no cracking and fracture at −15/−25 °C for hot melt type asphalt. A total of 50% of the data for SBS-modified asphalt was concentrated between −5 and −15 °C, which is lower than the specification requirement. The FYT waterproof coating and AMP-100 average value of −15 °C and −17 °C, respectively, can meet the needs of the specification 9 in the case of warm conditions. Waterborne epoxy emulsified asphalt and SBR-modified emulsified asphalt meet the requirements of specification 9 for waterborne asphalt materials under cold conditions. It shows that the waterborne epoxy emulsified asphalt and SBR-modified emulsified asphalt have excellent low-temperature performance among the water emulsion asphalt type materials. In addition, the low-temperature performance of epoxy bitumen is poor, with data mainly concentrated between −5 and −15 °C. In terms of impermeability, most materials meet the specification of 0.3 MPa pressure for 30 min without water penetration, but because the waterproofing bonding layer's resistance to construction damage is more important, it should be considered in conjunction with the study of the material's own resistance to water penetration. For the drying times, in the examples investigated, the surface and actual drying times were 3.5 h and 7.5 h for solvent asphalt, and 4 h and 8 h for both AMP-100 and waterborne epoxy emulsified asphalt, respectively.

To further ensure the long-term performance of the waterproofing binder material, the researchers also tested the corrosion resistance of the asphalt-based waterproof bonding material. Gao, Guo, Guo et al. studied the acid and alkali corrosion resistance of AMP-100, FYT, SBS-modified asphalt, and epoxy asphalt materials by immersing the materials in 2% H_2SO_4, 1% NaOH or 2% NaOH, and 20% NaCl solutions for 7~15 d. It was found that the four materials had good acid and alkali resistance [33–35]. Zhou immersed the epoxy asphalt specimens into 20% HCl, 20% NaOH, 20% NaCl, and 20% CH_3COONa solutions for 15 d, followed by tensile properties tests, respectively, and found that the epoxy asphalt specimens soaked in the solutions showed a small magnitude of mass change and a slight decrease in tensile strength and elongation at break [26].

3.2. *Investigation and Evaluation of the Bonding Performance of Asphalt-Based Waterproof Bonding Layers*

Due to poor shear resistance and complex construction processes, the coil-type waterproof bonding layer currently used in road engineering was relatively small. At the same time, temperature is the key factor affecting the performance of the waterproof bonding layer. Therefore, based on temperature, we have combed the research work of asphalt coating types and gravel sealed waterproof bonding layers for bridge deck pavements in China. The difference in adhesive properties of asphalt binder course materials is clarified.

3.2.1. Modified Asphalt Waterproof Bonding Layer Materials Bonding Properties

Bonding property statistics for epoxy asphalt at different temperatures can be seen in Figure 7. Bonding property statistics for SBS-modified asphalt can be seen in Figure 8. Rubber-modified asphalt bonding property statistics are shown in Figure 9. All data are from literature [5,9,10,13,22,26–50].

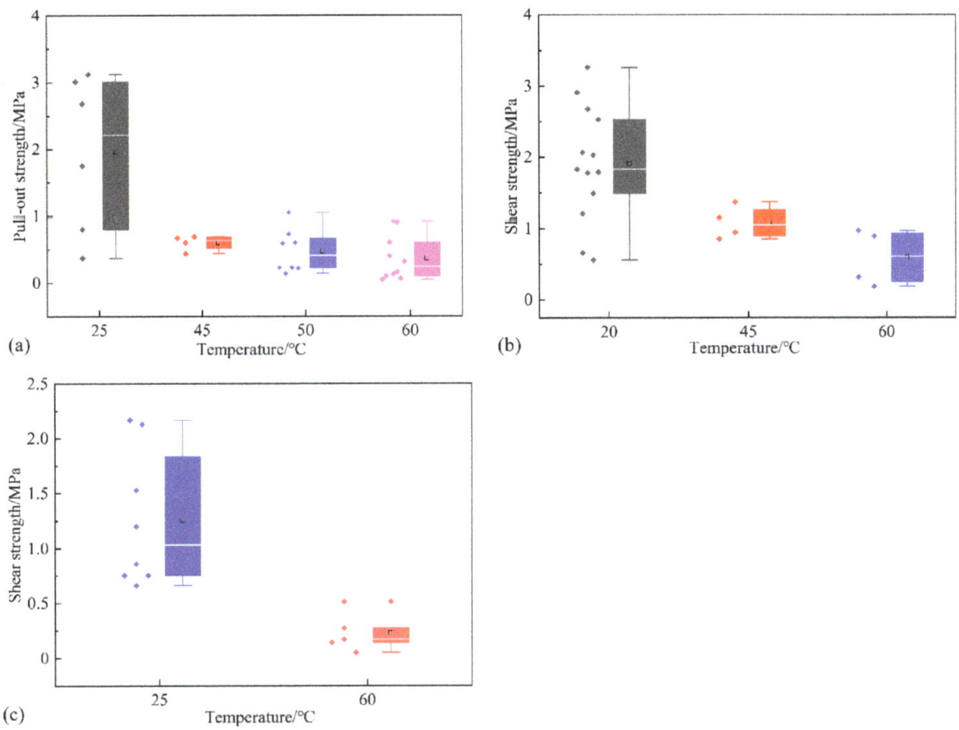

Figure 7. Bonding properties of the epoxy asphalt waterproof bonding layer materials. (**a**) Pull-out strength, (**b**) shear strength (direct shear), and (**c**) shear strength (40° oblique shear).

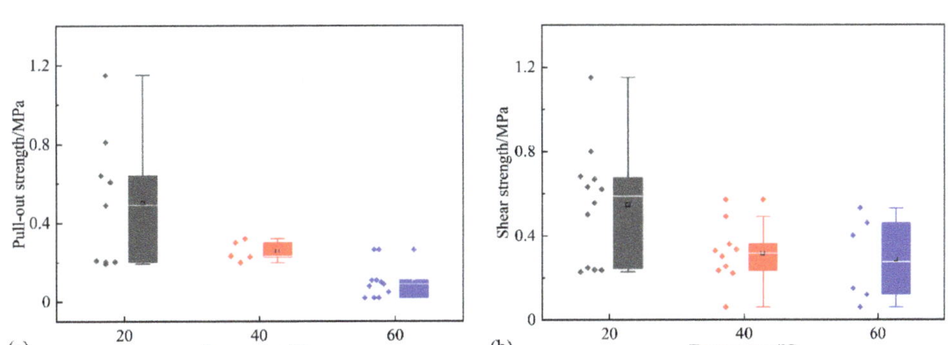

Figure 8. *Cont.*

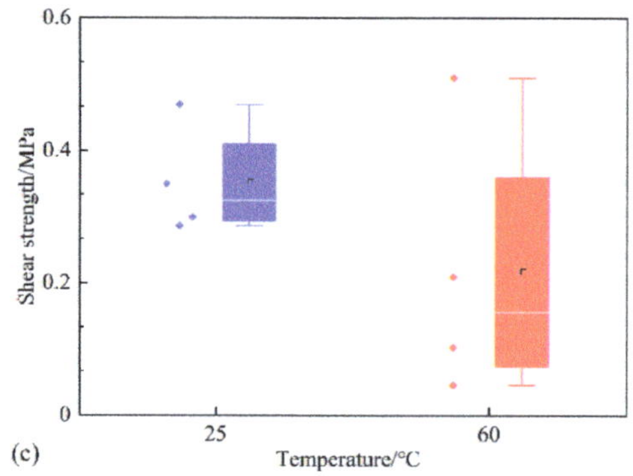

Figure 8. Bonding the properties of SBS-modified asphalt waterproof bonding layer materials. (**a**) Pull-out strength, (**b**) shear strength (direct shear), and (**c**) shear strength (45° oblique shear).

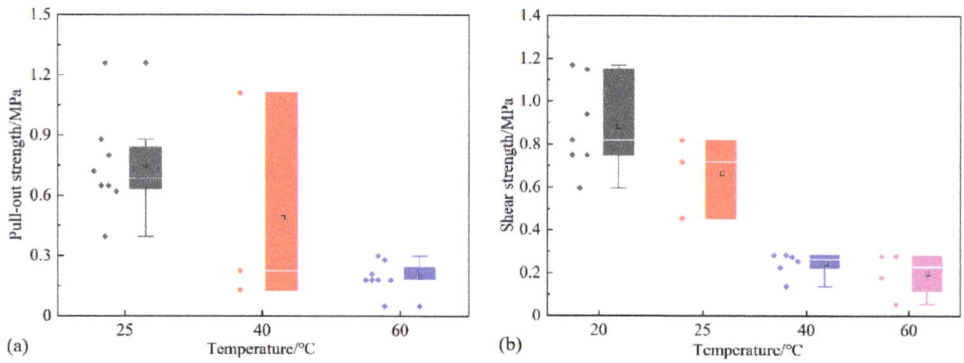

Figure 9. Bonding properties of the rubber-modified asphalt waterproof bonding layer materials. (**a**) Pull-out strength, and (**b**) shear strength (direct shear).

Epoxy asphalt is mainly applied to the coating of concrete bridge, making it a small part of the epoxy asphalt gravel waterproof bonding layer. In Figure 7, for the pull-out strength, which is highest at 25 °C, the mean value is approximately 1.96 MPa, meeting the specification of above 1.0 MPa. As the temperature rises, the overall trend of interlaminar drawing strength decreases, remaining at around 0.6 MPa at 45 °C, with an average value of only 0.36 MPa at 60 °C. For shear strength, the direct shear strength is mainly distributed between 1.49 and 2.53 MPa at 20 °C and remains at around 0.59 MPa at 60 °C. A total of 50% of the data for oblique shear strength at 25 °C are concentrated in the range of 0.75 to 1.83 MPa, while the average value at 60 °C is only 0.23 MPa. The pattern is basically the same as that for drawing strength, which decreases with increasing temperature. It can be seen that even though the asphalt and epoxy resin form a highly cross-linked mesh structure to improve strength and high temperature properties, as a temperature sensitive material, there is a certain degradation of mechanical properties when the temperature increases.

At present, the SBS-modified asphalt bonding material in the actual project is used with a single particle size of gravel together with the spread of a SBS-modified asphalt synchronous gravel waterproof bonding layer. Compared to coated materials, this structure not only has a good ability to coordinate deformation, but also effectively reduces construction costs and greatly improves the controllability of construction quality. In Figure 8, for the pull-out strength, the mean value is as high as 0.50 MPa at 20 °C. The pull-out strength decreases significantly with increasing temperature and only remains at 0.10 MPa at 60 °C.

As the median value is lower than the mean value, this indicates that the overall bonding performance is low at high temperatures. The shear strength is consistent with the above pattern, with the direct shear strength concentrated between 0.24–0.67 MPa at 20 °C, with a lower variation during the rise from 40 °C to 60 °C. The 45° oblique shear strength is mainly distributed between 0.29 and 0.41 MPa at 20 °C and concentrated between 0.07 and 0.36 MPa at 60 °C. In fact, when the asphalt surface paving uses a SBS-modified asphalt bonding material in the role of high temperature melting, the excess part is squeezed into the pores of the surface layer, and the surface layer shares the role of waterproofing, but the process is mainly physical bonding, and thermal stability is poor, as is the bonding layer of high-temperature adhesion performance in general.

Analysis of Figure 9 shows that the pull-out strength of the rubber-modified asphalt bonding material at 25 °C was mainly distributed in the range of 0.64–0.84 MPa, and as more concentrated in the range of 0.64~0.69 MPa. With the increase in temperature, pull-out strength decreases significantly, as the average value of strength at 60 °C is only 26.7% of that at 25 °C, which is about 0.20 MPa. The shear strength also decreases with increasing temperature, while the overall shear strength at 40 °C and 60 °C is relatively similar, with the former having a mean value of 0.24 MPa and the latter a mean value of 0.20 MPa. As the rubber particles become slightly loose and soft at high temperatures, the particles regain a certain degree of plasticity and adhesion, allowing them to maintain good adhesion at high temperatures.

3.2.2. Water Emulsion Type Asphalt Waterproof Bonding Layer Materials Bonding Properties

Statistics on the bonding properties of waterborne epoxy emulsified asphalt at different temperatures [10,32,51–57] were shown in Figure 10. Binding property statistics for SBS-modified emulsified asphalt [26,40,42,47,48,50] can be seen in Figure 11. The SBR-modified emulsified asphalt bonding property statistics [13,22,40,58,59] can be seen in Figure 12. The FYT bonding property statistics [13,24,25,34,60–62] can be seen in Figure 13. The AWP and AMP waterproofing adhesive material bonding property statistics [25,27,29,30,62,63] see Figure 14. In addition, some data are from literature [64–76].

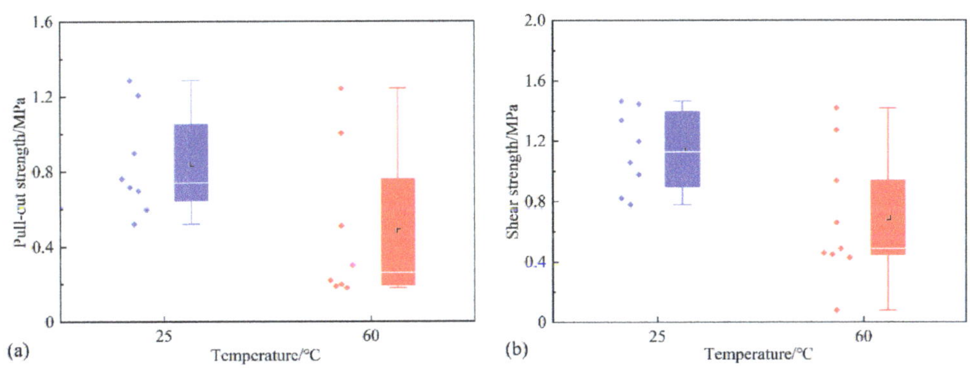

Figure 10. Bonding properties of the waterborne epoxy emulsified asphalt waterproof bonding layer materials. (**a**) Pull-out strength; (**b**) shear strength (direct shear).

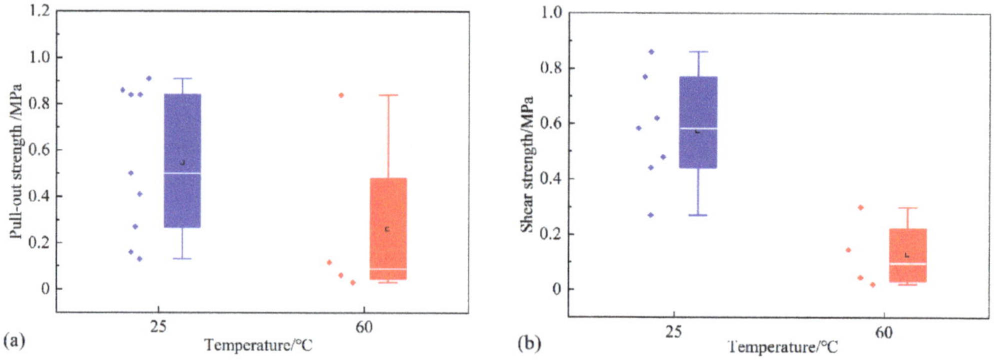

Figure 11. Bonding properties of the SBS-modified emulsified asphalt waterproof bonding layer materials. (**a**) Pull strength; (**b**) Shear strength (Direct shear).

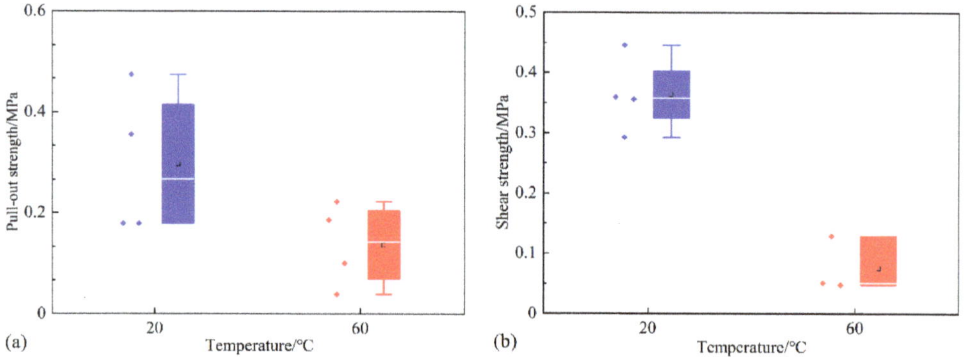

Figure 12. Bonding properties of the SBR-modified emulsified asphalt waterproof bonding layer materials. (**a**) Pull-out strength; (**b**) shear strength (direct shear).

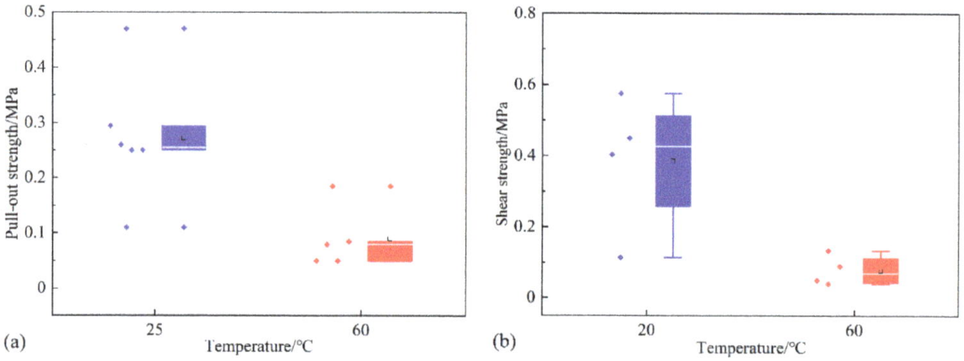

Figure 13. Bonding properties of the FYT waterproof bonding layer. (**a**) Pull-out strength; (**b**) shear strength (direct shear).

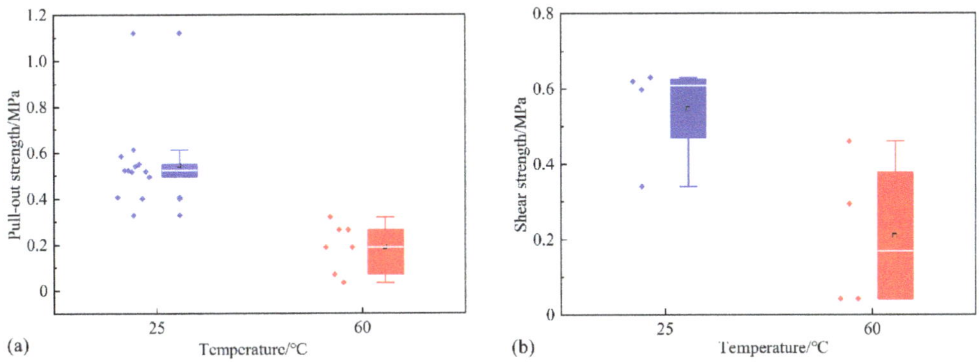

Figure 14. Bonding properties of the AWP and AMP waterproof bonding layer. (**a**) Pull strength; (**b**) shear strength (direct shear).

In Figure 10, for the pull-out strength, it was mainly concentrated between 0.65~1.05 MPa at 25 °C and maintained at 0.48 MPa at 60 °C. Among these values, a few data have higher pull-out strength at high temperature, which is mainly due to the higher admixture of epoxy resin; the epoxy resin can form more reticulation with asphalt to improve the strength, or the admixture of other viscosity increasing modified materials can also play a role. For shear strength, the mean value of direct shear strength at 25 °C is approximately 1.14 MPa, compared to 0.69 MPa at 60 °C. The data was more concentrated in the range of 0.45 to 0.49 MPa, but there are still some materials with good high temperature properties.

Analysis of Figure 11 shows that the pull-out strength of the SBS-modified emulsified asphalt bonding material at 25 °C was mainly concentrated in the range of 0.27–0.84 MPa, with a high degree of data dispersion (not concentrated). At 60 °C, it mainly remains at 0.26 MPa, but its median is only 0.09 MPa, which shows a high degree of overall attenuation of its pull-out strength at high temperatures. At 25 °C, the shear strength was mainly distributed between 0.44 and 0.77 MPa, and at 60 °C there is also a significant reduction, with the shear strength remaining at only 0.13 MPa.

Analysis of Figure 12 shows that, at 20 °C, the SBR-modified emulsified asphalt waterproof bonding layer's pull-out strength was mainly distributed between 0.18~0.41 MPa, and the data was more scattered. At 60 °C, when the strength dropped significantly, the data were mainly distributed between 0.07~0.20 MPa. A shear strength at 20 °C was mainly maintained at 0.36 MPa. At 60 °C when the strength is low, its average value is only 0.08 MPa, compared with at 20 °C. The mean value of strength decreased by about 77.8% compared with 20 °C. It can be seen that although SBR can improve the viscosity and toughness of asphalt, its high temperature stability is poor and it is difficult to apply in high temperature areas.

Analysis of Figure 13 shows that, at 25 °C, the FYT waterproof coating's pull-out strength was mainly concentrated between 0.25~0.29 MPa. At 60 °C, when the pull-out strength was mainly distributed in 0.05~0.09 MPa, the strength is relatively low. Shear strength versus the temperature is decreasing trend, sd its strength at 25 °C is 0.39 MPa but, at 60 °C, it reduced to 0.08 MPa. Thus, it can be seen that the high temperature performance of general water emulsion asphalt materials is poor.

The AWP and AMP waterproof coating was mainly based on asphalt, with rubber, resin, and other polymer materials used in the prepared water-based waterproof coating. It uses a polymer emulsion as the main film-forming substances, which include the main types of AWP-2000, AWP-2000F fiber-reinforced waterproofing materials and AMP-100, etc. Among them, AMP-100 has a unique first- and second-order reaction process, and construction can facilitate a certain degree of repair to concrete bridge deck micro-cracks. After curing, they can form a highly elastic plastic coating film. In addition, the material and asphalt surface compatibility is good, the bonding ability, and waterproofing effect are relatively strong, and the AWP-2000 type in the coating after curing can have a certain degree of elasticity, allowing it to play a stress absorption role. In Figure 14, the pull-out strength is mainly maintained at 0.55 MPa at 25 °C and decreases to 0.19 Mpa at 60 °C; the shear strength is consistent with the overall pattern of pull-out strength, with the average value decreasing from 0.55 Mpa at 25 °C to 0.21 Mpa at 60 °C.

In addition to the above commonly used types, researchers have also conducted studies on other types of modified emulsified asphalt, such as CR and EVA and modified emulsified asphalt thin slurry sealers, etc. Therefore, statistical summaries of the bonding performance of other types of modified emulsified asphalt waterproofing binder materials other than those mentioned above [14,27,40,55,60,64–66] are shown in Figure 15.

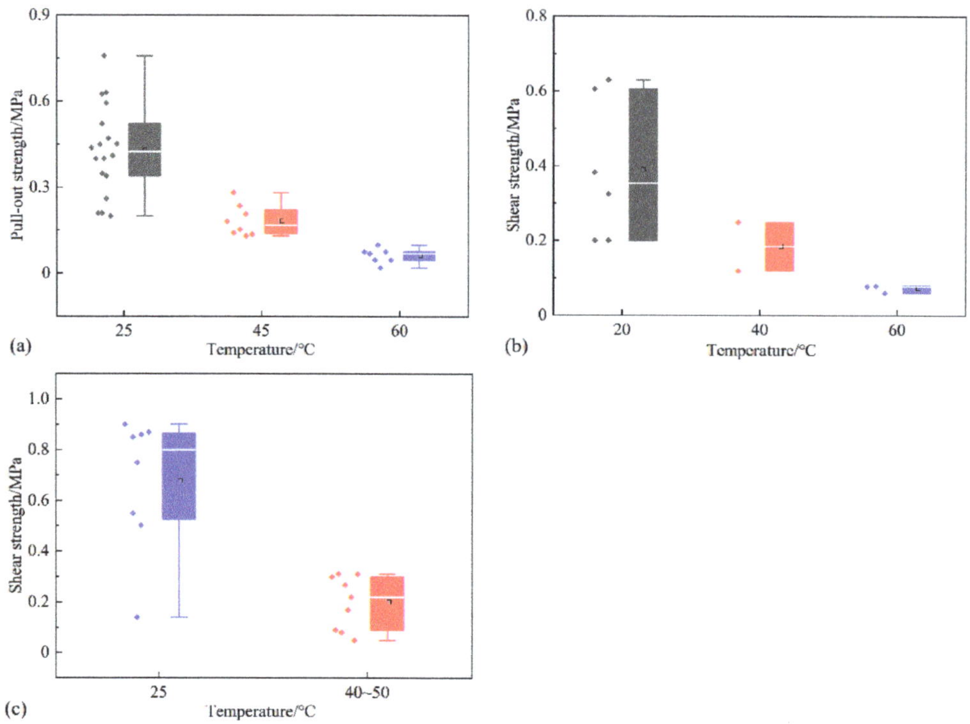

Figure 15. Bonding properties of other types of modified emulsified asphalt waterproof bonding layer materials. (**a**) Pull strength; (**b**) shear strength (direct shear); (**c**) shear strength (40° oblique shear).

Analysis of Figure 15 shows that for the drawing strength, the most data were collected at 25 °C and were mainly distributed in the range of 0.34~0.52 MPa, with the temperature increasing against the overall decreasing trend of strength. The average value was about 0.18 MPa at 45 °C and was only maintained at 0.06 MPa at 60 °C; the pattern of interlayer shear strength and drawing strength is basically the same, as the most data were collected at 20 °C for the direct shear strength, but its dispersion was high. The degree of the shear angle was higher with the temperature of 60 °C, as it was only 0.07 MPa; at the shear angle of 40 °, the average value of shear strength from 25 °C at 0.68 MPa decreased to 40~50 °C at 0.20 MPa. Overall, the modified emulsified asphalt waterproofing binder layer offers a low performance at high temperatures.

3.2.3. Solvent Asphalt Waterproof Bonding Layer Materials Bonding Properties

Solvent asphalt was generally not used alone as a waterproof bonding layer, but with epoxy resin and rubber asphalt sand adhesive together to form a waterproof bonding system. Its bonding performance statistics [23,27,28,31,67–69,75,76] can be seen in Figure 16.

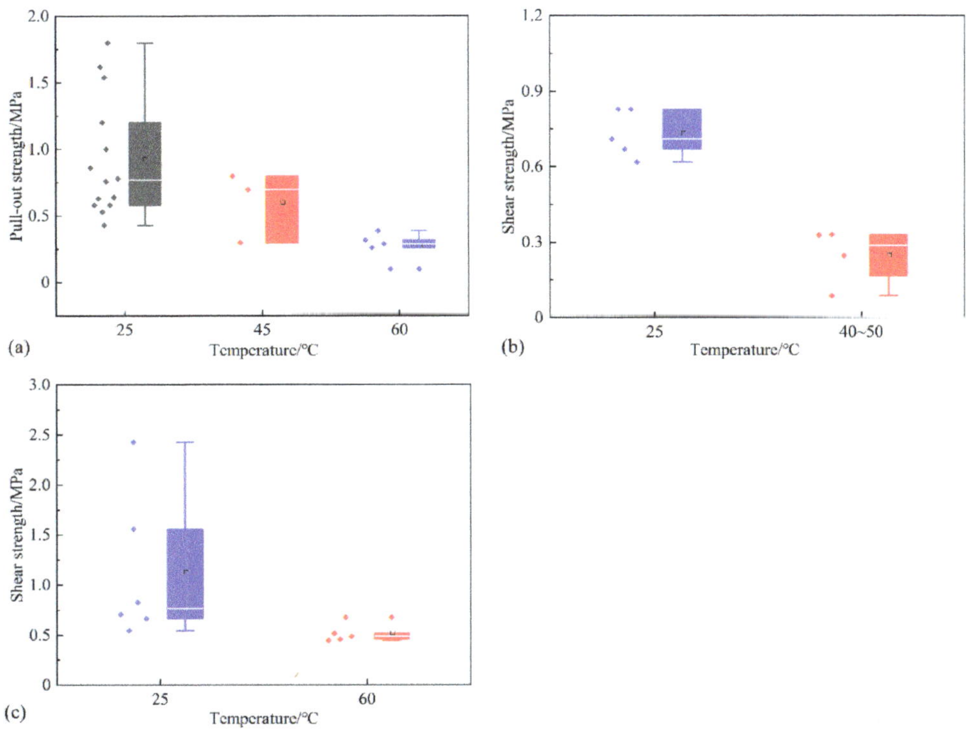

Figure 16. The bonding properties of solvent asphalt waterproof bonding layer materials (**a**) Pull strength; (**b**) shear strength (direct shear); (**c**) shear strength (40° oblique shear).

Analysis of Figure 16 shows that the most data were collected at 25 °C, and the pull-out strength was mainly distributed in the range of 0.58~1.20 MPa, with a high degree of data dispersion, and a few parts were higher than 1.5 MPa, which was mainly due to the improvement of interlayer bonding performance by setting epoxy resin as the under-seal layer. The pull-out strength showed a decreasing trend as the temperature rose, maintaining at 0.6 MPa at 45 °C, and mainly concentrated between 0.26~0.32 MPa at 60 °C. Shear strength also decreased with the rise in temperature, as direct shear strength at 25 °C is mainly concentrated between 0.67~0.83 MPa, while at 40~50 °C it is at 0.25 MPa. With a shear angle of 40 °, the shear strength changes from 1.13 MPa at 25 °C to 0.52 MPa at 60 °C. In addition, by including the rubber asphalt sand rubber, it can also play a role in protecting the waterproof bonding layer and absorbing stress.

3.3. Comparison of Bonding Properties of Different Asphalt-Based Waterproof Bonding Layers

According to the data from Figures 7–16, a comparative analysis of different types of waterproof bonding layer bonding performance indicators can be determined. Specific indicators include 25 °C and 60 °C pull-out strength, and shear strength (direct shear), as can be seen in Figures 17 and 18.

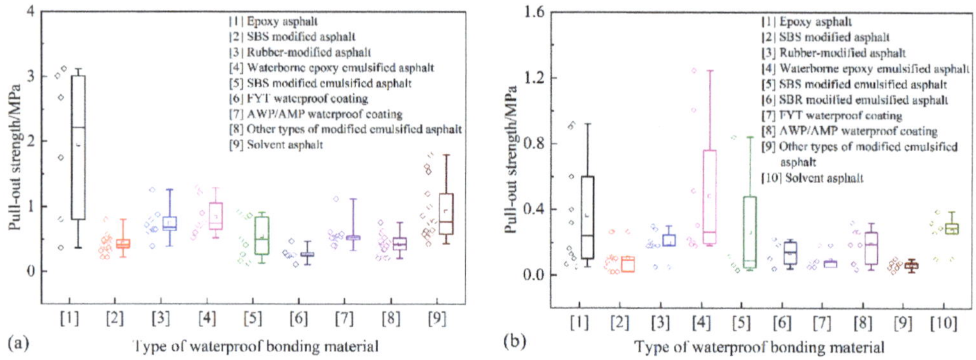

Figure 17. Comparison of pull strength of different asphalt waterproof bond layers. (**a**) Pull-out strength (25 °C); (**b**) pull-out strength (60 °C).

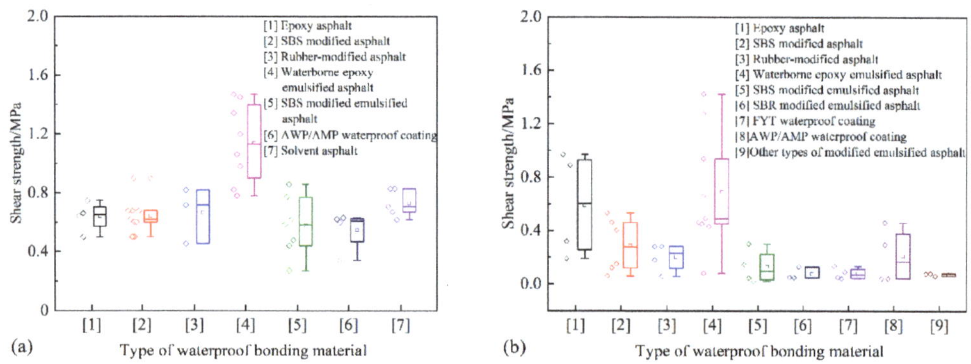

Figure 18. Comparison of shear strength of different asphalt waterproof bond layers. (**a**) Shear strength (direct shear 25 °C); (**b**) shear strength (direct shear 60 °C).

Analysis of Figure 17a shows that the epoxy asphalt mean and median are highest, followed by the solvent asphalt waterproofing system, then the waterborne epoxy emulsified asphalt, indicating that the epoxy resin and solvent-based asphalt constitute a composite structure which can greatly enhance the adhesion and flexibility of the material. Rubber-modified asphalt synchronous gravel waterproofing bonding layer has a high interlayer pull-out strength, and is overall higher than SBS-modified asphalt, mainly due to the fact that the overall amount of SBS-modified asphalt is less, and the gravel spreading bonding material coating is not uniform, resulting in more voids which affect the bond strength. In addition, it can be seen that of the water emulsion asphalt materials, except in the analysis of Figure 17a, epoxy asphalt has the highest mean and median, followed by the solvent asphalt waterproofing system, then the waterborne epoxy emulsified asphalt, indicating that the epoxy resin and solvent asphalt constitute a composite structure, which can greatly enhance the adhesion and flexibility of the material. Rubber-modified asphalt synchronous gravel waterproof bonding layer has a high interlayer pull-out strength, and is overall higher than SBS-modified asphalt, mainly due to the fact that the overall amount of SBS-modified asphalt is less, and the gravel spreading bonding material coating is not uniform, resulting in more voids which affect the bond strength. In addition, for water emulsion asphalt materials, except waterborne epoxy emulsified asphalt, the overall pull-out strength is low. This includes FYT and other types of modified emulsified asphalt bonding, meaning that the performance is poor. This is mainly because the general water emulsion asphalt materials contain about 40% water, so that the formation of the thin asphalt film after curing affects the bonding performance. Analysis of Figure 17b shows that as asphalt is by nature a highly temperature sensitive viscoelastic material, the properties will decay to a certain extent at high temperatures, resulting in a higher decay in pull-out strength at 60 °C compared to 25 °C for different asphalt materials. The average pull-out strength of waterborne epoxy emulsified asphalt is the highest, at 0.48 MPa, followed by epoxy asphalt, solvent asphalt, and SBS-modified emulsified asphalt, at 0.37 MPa, 0.27 MPa, and 0.26 MPa, respectively, while the pull-out strength of rubber-modified asphalt binder materials is mainly concentrated between 0.18~0.25 MPa, lower than the above four categories. The other water emulsion asphalt, AWP and AMP, have a relatively high waterproof coating performance, as its pull-out strength is maintained at 0.19 MPa, while FYT and other types of modified emulsified asphalt can only be maintained at 0.10 MPa, 0.09 MPa and 0.06 MPa.

From Figure 18a it can be seen, the overall shear strength from high to low for waterborne epoxy emulsified asphalt, solvent asphalt, epoxy asphalt, rubber-modified asphalt, SBS-modified asphalt, SBS-modified emulsified asphalt and AWP and AMP are relatively similar, except for waterborne epoxy emulsified asphalt and solvent asphalt. Analysis of Figure 18b shows that waterborne epoxy emulsified asphalt and epoxy asphalt has the best high temperature shear strength, followed by SBS-modified asphalt and rubber-modified asphalt, meaning that a synchronous gravel waterproofing bonding layer at high temperatures offers good coordination for shear deformation resistance. The remaining materials in the AWP and AMP waterproof coating offer relatively good high temperature performance, but its cost is higher.

The epoxy asphalt, waterborne epoxy emulsified asphalt, and solvent asphalt waterproof bonding system have the best interlayer bonding performance, because the epoxy resin has good mechanical properties, so that the three-dimensional interpenetrating mesh structure formed with the asphalt, which fundamentally changed the ordinary asphalt thermoplastic, greatly enhancing the high temperature stability of the material [77]. In addition, waterborne epoxy emulsified asphalt can be used for cement concrete bridge decks to repair micro-cracks during the construction period due to its good fluidity and permeability. The SBS-modified asphalt and rubber-modified asphalt synchronous gravel waterproof bonding layer with the advatanges of convenient construction, quality control, low cost and good interlayer shear resistance and other advantages were widely paid attention to by researchers. However, compared with the coating class materials, its high internal void ratio leads to low bonding performance and its waterproofing performance is average [78]. The reason for this is that it is difficult to form an effective embedding structure after spreading single-grain gravel, which makes it easy for the gravel to be incompletely coated during construction, resulting in a high porosity and susceptibility to water infiltration and ponding. The water emulsified asphalt, when used as a coating material, has average interlayer bonding properties. The subsequent study can be improved by mixing the appropriate grade of coarse and fine aggregates to improve the interlayer shear resistance. In addition, as the emulsified asphalt has an electrical charge on its surface, the asphalt particles can be closely adsorbed to the aggregate surface, which not only reduces the void ratio but also increases the adhesion to the aggregate, giving it a good waterproofing effect.

3.4. Construction Process of Asphalt-Based Waterproof Bonding Layer Materials

The construction process of waterproof bonding materials is one of the important factors affecting the quality of the project. At present, there are many different types of adhesive materials and different bonding principles and applicable conditions, resulting in different types of waterproof bonding layer construction process differences. According to existing research results [79–85], the construction process of the waterproof bonding layer can be distinguished by coil, coating, and a gravel sealing layer. The thickness of the coiled material is high, and the construction is mostly carried out by manual on-site baking. After the asphalt is melted and rolled, it is bonded to the concrete panel. The coating materials are constructed through an asphalt distributor, an intelligent distributor, mechanical spraying, and manual scraping. Among these, SBS-modified asphalt, SBR-modified emulsified asphalt, and other materials are sprayed on-site by asphalt distributors. The temperature, speed, and dosage of asphalt spraying should be kept stable and artificial spraying should be carried out in time. Epoxy asphalt and waterborne epoxy emulsified asphalt are generally spread by intelligent spreaders to ensure uniform, continuous, and accurate dosage. To ensure uniformity, continuity, and accurate dosage, the binder materials, such as AMP-100, AWP-2000, and solvent-type asphalt are applied by mechanical spraying or manual scraping, etc. To avoid holes in the drying process of the binder materials, the construction is generally applied in two layers. Synchronous gravel sealer is mostly constructed by a synchronous gravel sealer truck, which should be driven smoothly and evenly and that ensure the temperature of asphalt sprinkling is between 160~180 °C. After the synchronous gravel sealer is sprinkled, it is rolled by rubber roller in time.

It is worth noting that there is a significant difference between the construction interval time of different asphalt binder materials. The upper surface layer construction can be carried out after the completion of the bonding, and the interval time is almost zero. The construction interval of heat-modified asphalt bonding materials and a synchronous crushed stone seal is relatively short. The SBS- and SBR-modified emulsified asphalt and other water emulsion materials should be carried out on the upper surface layer after the demulsification of emulsified asphalt and the complete evaporation of water, and the interval time is about 8~4 h. The drying time of epoxy and waterborne epoxy emulsified asphalt binder should be controlled between 3~5 h. The interval between the first and second coating construction of AMP-100 is 2~8 h, and the asphalt surface layer construction is carried out after the complete drying. The solvent-based asphalt is the second layer coating after the first layer is dry. In addition, the coil has higher requirements on the flatness of the bridge deck, which needs to be baked while rolling; synchronous crushed stone seals are prone to incomplete gravel coating during construction. The coating rate can be improved by spraying asphalt twice before and after gravel spreading.

4. Conclusions and Prospects

(1) For the asphalt waterproof bonding layer, the epoxy asphalt waterproof adhesive material interlayer bonding has the best performance. The waterborne epoxy emulsified asphalt has good high-temperature performance and excellent fluidity and permeability. The SBS and rubber-modified asphalt synchronous crushed stone seal has the advantages of convenient construction, controllable quality, low cost, and good high-temperature coordination shear resistance.

(2) Key performance indicators of asphalt-based waterproof bonding layers material include heat resistance, low-temperature flexibility, impermeability, and bonding.

(3) Epoxy asphalt, solvent asphalt, and waterborne epoxy emulsified asphalt bonding performances are high, with the average value of pull-out strength of 1.96, 0.92, 0.84 MPa, respectively. Epoxy asphalt, waterborne epoxy emulsified asphalt, and solvent-based asphalt waterproof bonding systems have the best high-temperature performance, followed by SBS-modified asphalt and a rubber-modified asphalt synchronous crushed stone waterproof bonding layer.

In the future, more in-depth research and exploration should be carried out in the following aspects: (1) Scientific and reasonable index evaluation systems and advanced test methods for the bridge waterproof adhesive layer should be further put forward; (2) Based on the existing engineering experience and research results, the performance evaluation system of the waterproof adhesive layer suitable for different regions and different material types should be established.

Author Contributions: Conceptualization, F.G., Q.C. and C.W.; methodology, Z.G.; validation, X.G.; formal analysis, F.G.; investigation, X.G.; resources, C.W.; data curation, Y.L.; writing—original draft preparation, F.G.; writing—review and editing, Q.C. and C.W.; supervision, X.G.; project administration, C.W. and Q.C. All authors have read and agreed to the published version of the manuscript.

Funding: This research was sponsored by the Innovation Capability Support Program of Shaanxi (2022TD-07).

Data Availability Statement: All data, models, and code generated or used during the study appear in the published article.

Conflicts of Interest: The authors declare no conflict of interest.

References

1. Fu, H.; Wang, C.; Niu, L.; Yang, G.; Liu, L. Composition optimisation and performance evaluation of waterborne epoxy resin emulsified asphalt tack coat binder for pavement. *Int. J. Pavement Eng.* **2021**, *2021*, 1932878. [CrossRef]
2. Liu, L.Q.; Wang, C.H.; Liang, Q. Preparation of a heat insulation bonding layer for roads and its heat insulation effect. *J. Clean. Prod.* **2022**, *365*, 132828. [CrossRef]
3. Chen, Q.; Wang, C.; Hu, X.; Guo, C.; Liu, F. Preparation and property optimization of road basic energy-absorbing materials based on balanced control. *Acta Mater. Compos. Sin.* **2022**, *39*, 3356–3368.
4. Haynes, M.A.; Coleri, E.; Sreedlhar, S. Impermeable Asphalt Concrete Layer to Protect and Seal Concrete Bridge Decks. *Transp. Res. Rec.* **2019**, *2673*, 355–367. [CrossRef]
5. Kang, H.; Oh, K.; Ahn, K.; Jiang, B.; He, X.; Oh, S. Property Analysis of Double-Sided Composite Waterproofing Sheet for Simultaneous Application on Asphalt Concrete and Latex-Modified Concrete Pavements for Bridge Decks. *Appl. Sci.* **2022**, *12*, 9779. [CrossRef]
6. Mazzotta, F.; Lantieri, C.; Vignali, V.; Simone, A.; Dondi, G.; Sangiorgi, C. Performance evaluation of recycled rubber waterproofing bituminous membranes for concrete bridge decks and other surfaces. *Constr. Build. Mater.* **2017**, *136*, 524–532. [CrossRef]
7. Oh, K.H.; Kim, S.Y.; Park, Y.G. Joint- or Crack-Opening Resistance Evaluation of Waterproofing Material and System for Structural Sustainability in Railroad Bridge Deck. *Materials* **2020**, *13*, 4229. [CrossRef]
8. Zhang, F.; Li, M.Q.; Wang, T.Y.; Feng, D.C. Performance of composite waterproof cohesive layer on cement concrete bridge. *J. Harbin Inst. Technol.* **2020**, *52*, 26–32.
9. Wan, C.G.; Shen, A.Q.; Zhao, X.Y. Xun, C. Grey target optimization of waterproof adhesive materials for bridge deck pavement with comprehensive. *J. Build. Mater.* **2017**, *20*, 406–410.
10. Hailesilassie, B.W.; Hean, S.; Partl, M.N. Testing of blister propagation and peeling of orthotropic bituminous waterproofing membranes. *Mater. Struct.* **2015**, *48*, 1095–1108. [CrossRef]
11. Chen, Q.; Wang, C.H.; Fu, H.; Yang, G.; Sun, Z. Optimization of construction method of waterborne epoxy asphalt cape seal based on performance evolution. *China J. Highw. Transp.* **2021**, *34*, 236–245.
12. Zhang, H.; Gao, P.; Zhang, Z.; Pan, Y. Experimental study of the performance of a stress-absorbing waterproof layer for use in asphalt pavements on bridge decks. *Constr. Build. Mater.* **2020**, *254*, 119290. [CrossRef]
13. Liu, M.M.; Han, S.; Yang, H.; Wu, X. The applicability of different waterproof adhesive interlayer for bridge deck pavement. *J. Shenzhen Univ. (Sci. Eng.)* **2020**, *37*, 103–110. [CrossRef]
14. Editorial Department of China Journal of Highway and Transport. Review on China's bridge engineering research: 2014. *China J. Highw. Transp.* **2014**, *27*, 1–96.
15. Huang, X.M. Research status summary of asphalt pavement technology on cement concrete bridge deck. *J. Traffic Transp. Eng.* **2014**, *14*, 1–10.
16. Chen, Q.; Wang, C.H.; Yu, S.X.; Song, Z.; Fu, H.; An, T. Low-temperature mechanical properties of polyurethane-modified waterborne epoxy resin for pavement coating. *Int. J. Pavement Eng.* **2022**, 1–13. [CrossRef]
17. Wang, C.H.; Shu, C.; Han, B.; Fu, Y.; Zhou, L. Research progress of high modulus asphalt concrete. *J. Chang. Univ. (Nat. Sci. Ed.)* **2020**, *40*, 1–15.
18. *DB32/T 2285-2012*; Waterborne Epoxy-Binder Waterproof Coating for Concrete Bridges. Jiangsu Provincial Bureau of Quality and Technical Supervision: Nanjing China, 2012.
19. *JC/T 408-2005*; Emulsified Asphalt Waterproof Coating. China Chemical Building Materials Co., Ltd.; Suzhou Waterproof Material Research and Design Institute; National Development and Reform Commission of the People's Republic of China: Beijing, China, 2005.
20. Chen, Z.Y. Study on the Waterproof Binding Layer of Long Span Orthogonal Intersection Steel Bridge Deck. Master's Thesis, Chang'an University, Xi'an, China, 2008.
21. Liu, Z.Y. Research on Concrete Bridge Deck Waterproofing in Cold Regions. Master's Thesis, Chang'an University, Xi'an, China, 2008.
22. Zhang, J. Study on the Waterproofing and Bonding Layer of Concrete Bridge Deck. Master's Thesis, Chang'an University, Xi'an, China, 2008.
23. Zhou, J.W. Experimental Study on Waterproof Binding Course of Concrete Bridge Deck. Master's Thesis, Nanjing Forestry University, Nanjing, China, 2009.

24. Cai, Y.X. Study on Composite Bridge Deck Waterproof Adhesive Layer for Humid and Rainy Area. Master's Thesis, Chang'an University, Xi'an, China, 2011.
25. Zhou, B. Epoxy-Based Waterproof and Adhesive Materials Development for Concrete Bridge Deck. Master's Thesis, Chang'an University, Xi'an, China, 2012.
26. Yang, Y. Waterproofing Adhesive Material of Expressway Concrete Bridge Deck and Performance Study in East of Gansu Province. Master's Thesis, Chongqing Jiaotong University, Chongqing, China, 2012.
27. Zhang, D.L. Deck Assessment and Application of Waterproof and Adhesive Layer Material Properties. Master's Thesis, South China University of Technology, Guangzhou, China, 2016.
28. Guo, J.B. Cement Concrete Bridge Deck Waterproofing Adhesive Layer Performance Research and Application. Master's Thesis, South China University of Technology, Guangzhou, China, 2014.
29. Shi, C.G. Research on Performance of Bridge Deck Material Structure Combination and Waterproof Adhesive Layer. Master's Thesis, Jilin University, Jilin, China, 2015.
30. Huang, X.X. The Applied Research of Waterproof Bonding Layer of Asphalt Rubber in the Highway of Tanluo to Baise in Guangxi. Master's Thesis, Guangxi University, Xining, China, 2012.
31. Li, A.F. Research on the Application of New Water-Based Epoxy Waterproof Adhesive Layer in Cement Concrete Bridge Deck Pavement in Guizhou Mountain Area. Master's Thesis, Southeast University, Nanjing, China, 2016.
32. Gao, X.C.; Huang, X.M.; Wang, S.G. Research on waterproof and cohesive layer for long-span prestressed concrete bridge desk. *J. Highw. Transp. Res. Dev.* **2005**, *8*, 70–73.
33. Guo, Z.; Zhou, Z.G.; Deng, C.Q.; Yu, W.; Luo, G. Performance analysis of waterproof bonding layer of asphalt pavement on different cement concrete bridge deck. *Highway* **2018**, *63*, 87–93.
34. Guo, P. Research on Water-Proof Layer in Concrete Bridge Deck. Master's Thesis, Chang'an University, Xi'an, China, 2012.
35. Zhang, Z.Q.; Lei, Z.J.; Yang, B. Study on the performance of waterproof bonding layer of glass fiber asphalt bridge deck. *Highway* **2011**, *9*, 34–37.
36. Wang, J.; Zhou, B.; Chen, S.F.; Wang, J.; Lei, Y. Study on road performance of domestic epoxy asphalt bridge deck waterproof bonding material. *Highway* **2012**, *7*, 35–38.
37. Su, R.R.; Song, Z.J.; Song, X.D. Experimental research on waterproof binding course of epoxy asphalt deck. *Road Mach. Constr. Mech.* **2015**, *32*, 84–86+91.
38. Zhang, Z.Q.; Tao, J.; Zhang, S.T. Experiment and evaluation on performance of epoxy asphalt waterproof cohesive layer on bridge deck pavement. *J. Chang. Univ. (Nat. Sci. Ed.)* **2011**, *31*, 1–6.
39. Ma, Y.Q.; Liang, Z.L. Study on the selection and properties of bonding waterproof layer materials for bridge deck pavement. *Highway* **2007**, *6*, 101–103.
40. Zhang, J. Shear Performance of Waterproofing and Bonding Layer for Concrete Bridge Deck. *J. Highw. Transp. Res. Dev.* **2011**, *28*, 29–34.
41. Yu, X.; Li, Y.T.; Liu, Y. Grey Target Optimization of Waterproof Adhesive Material for Concrete Bridge Deck Pavement Based on Mechanical Analysis. *J. Southwest Jiaotong Univ.* **2012**, *47*, 1086–1091.
42. Wan, C.G.; Shen, A.Q.; Fan, L. Composite behavier of waterproof binding course on cement concrete bridge deck. *J. Hefei Univ. Technol. (Nat. Sci.)* **2016**, *39*, 960–964.
43. Hou, D.H.; Han, M.Z.; Muhammad, Y.; Liu, Y.; Zhang, F.; Yin, Y.; Duan, S.; Li, L. Performance evaluation of modified asphalt based trackless tack coat materials. *Constr. Build. Mater.* **2018**, *165*, 385–394. [CrossRef]
44. Li, Z.D.; Huang, X.M.; Chen, G.X.; Chen, X. On Modified Temperature Model for Synchronous Chip Seal Waterproof Binder Course on the Concrete Bridge. *J. Chongqing Jiaotong Univ. (Nat. Sci.)* **2011**, *30*, 957–964.
45. Yang, Y.S.; Li, Z.X.; Wang, R.D.; Gao, Z. Technology Research on "Two Asphalt One Aggregate" Waterproof Based on Synchronous Crushed Stone. *J. Hebei Univ. Technol.* **2010**, *39*, 92–95.
46. Yang, Y.S.; Li, Z.X.; Wang, X.C. Road performance of synchronous crushed stone waterproof binding course of bridge pavement. *J. Chang. Univ. (Nat. Sci. Ed.)* **2009**, *29*, 19–23+58.
47. Duan, B.D.; Li, J.; Li, M.L.; Cao, D.; Chen, F. Research on material parameters of waterproof adhesive layer for porous asphalt pavement based on interlayer adhesive performance. *Highw. Eng.* **2019**, *44*, 45–49+137.
48. Cao, M.M.; Huang, W.Q.; Lu, Y.; Wu, M.; Li, Y. Test and Evaluation method of Interlaminar Shear Property of Composite Pavement. *J. Highw. Transp. Res. Dev.* **2018**, *35*, 40–48.
49. Bao, C.F.; Yuan, M.Y.; Yu, J.M.; Wang, G.; Wang, J. Study on performance evaluation of bridge deck waterproof bond layer based on interlaminar shear test. *Highway* **2016**, *61*, 212–217.
50. Liang, J.; Xu, H.; Guo, H.; Han, S.; Xia, J. Research on influence of water-based epoxy resin modifier on performance and structure of emulsified asphalt. *Bull. Chin. Ceram. Soc.* **2020**, *39*, 1998–2004.
51. Wang, Q.Z.; Ma, X.J.; Xu, D.W.; Yang, Z. Study on the performance of waterborne epoxy modified emulsified asphalt bridge deck adhesive layer. *Thermosetting Resin* **2020**, *35*, 59–63.
52. Chang, Y.T.; Chen, Z.D.; Niu, X.H.; Zhou, Z. Test of shear resistance of modified emulsified asphalt by waterborne epoxy resin. *J. Jiangsu Univ. (Nat. Sci. Ed.)* **2017**, *38*, 224–229. [CrossRef]
53. Yang, G.L.; Wang, C.H.; Fu, H.; Yan, Z.; Yin, W. Waterborne epoxy resin-polyurethane-emulsified asphalt: Preparation and properties. *J. Mater. Civ. Eng.* **2019**, *31*, 04019265.1–04019265.11. [CrossRef]

54. Wang, M.; Wu, X.L.; Li, B.; Xiao, L.; Peng, Z. Experimental Research on Waterproof Adhesive Layer Material of Concrete Bridge Slab. *Highw. Eng.* **2021**, *47*, 91–96.
55. Ji, J.; Liu, L.H.; Suo, Z.; Xu, Y.; Yang, S.; Xu, S. Performances of Micro-Surfacing with Waterborne Epoxy Resin Modified Emulsified Asphalt. *J. Chang. Univ. (Nat. Sci. Ed.)* **2017**, *37*, 23–30.
56. Chen, Q.; Lu, Y.F.; Wang, C.H.; Han, B.; Fu, H. Effect of raw material composition on the working performance of waterborne epoxy resin for road. *Int. J. Pavement Eng.* **2022**, *23*, 2380–2391. [CrossRef]
57. Ma, L. Study on material performance of high-permeability bridge deck waterproof adhesive layer. *New Build. Mater.* **2020**, *47*, 164–168.
58. Yi, X.J. A Weighted multi-objective decision model for selecting optimum waterproof material of cement concrete bridge. *J. China Foreign Highw.* **2020**, *40*, 298–302.
59. Wang, Y.L.; Zhou, Y.L.; Yao, A.L.; Zhang, Y. Test of shear and pull-off between asphalt and concrete on bridge deck pavement structure. *J. Chang. Univ. (Nat. Sci. Ed.)* **2009**, *29*, 15–18.
60. Liu, S.H. Application of FYT(coating) bridge deck waterproof layer on binbo expressway. *J. China Foreign Highw.* **2005**, *2*, 128–130.
61. Meng, C.C. Optimization of waterproof bond system for concrete deck pavement in rainy and damp-heat area. *Road Mach. Constr. Mech.* **2018**, *35*, 108–111.
62. Liu, L.L.; Li, Y.S.; Zong, J.Y.; Wu, C. Comparative study on the properties of waterproof binding materials for rigid and flexible composite pavement of steel bridge deck. *China Adhes.* **2021**, *30*, 50–53+58.
63. Guo, W.H.; Chen, F.; Li, Y.; Chi, H. Performance test on waterproof binding layer of deck pavement of prestress concrete box-girders bridge with corrugated steel webs. *J. Xihua Univ. (Nat. Sci. Ed.)* **2017**, *36*, 103 106.
64. Li, Q.L.; Tian, Y.; Zhang, M. Study on factors affecting adhesive strength of emulsified asphalt water-proof coating. *New Build. Mater.* **2019**, *46*, 96–98. [CrossRef]
65. Ma, T.; Huang, X.M.; Ju, H. Research on the Performance of Waterproof and Cohesive Layer. *J. Highw. Transp. Res. Dev.* **2007**, *1*, 43–46.
66. Jiang, S.W. Sensitivity analysis of interlayer performance of waterproof bond with different gravel size range. *J. China Foreign Highw.* **2017**, *37*, 290–293.
67. Zhao, Z.Y.; Chen, F.L.; Liu, G.P. Material selection and analysis of waterproof bonding layer for steel bridge deck pavement. *Highway* **2017**, *62*, 288–291.
68. Zhang, D.J.; Zhu, H.P.; Pan, Y.Q. Tests for waterproof binding layer of steel bridge deck pavement of Fourth Nanjing Changjiang River Bridge and the paving technique. *World Bridges* **2014**, *42*, 44–48.
69. Jin, W.; Zhao, Y.; Wang, W.; He, F. Performance evaluation and optimization of waterproof adhesive layer for concrete bridge deck in seasonal frozen region using AHP. *Adv. Mater. Sci. Eng.* **2021**, *2021*, 5555535. [CrossRef]
70. Liu, W.; Yan, K.; Ji, H. Bonding performance evaluation on WTR-APAO composite modified asphalt as waterproof adhesive layer for concrete bridge. *Constr. Build. Mater.* **2022**, *349*, 128667. [CrossRef]
71. Fen, Y.; Kong, Z.; Xiao-yang, J.; Lu, Q.; Yan, J. Evaluation of shear performance of flexible waterproof-adhesive layer in concrete bridge pavement based on grey correlation analysis. *Road Mater. Pavement Des.* **2009**, *10* (Suppl. S1), 349–360. [CrossRef]
72. Qian, G.; Li, S.; Yu, H.; Gong, X. Interlaminar bonding properties on cement concrete deck and phosphorous slag asphalt pavement. *Materials* **2019**, *12*, 1427. [CrossRef]
73. Zha, X.; Lv, R.; Hu, H.; Chen, H.; Wang, Z. Preparation and experimental research on the properties of neodymium iron boron magnetic powder modified asphalt. *Constr. Build. Mater.* **2022**, *321*, 126392. [CrossRef]
74. Sun, Y.; Liu, Y.; Gong, J.; Han, X.; Xi, Z.; Zhang, J.; Wang, Q.; Xie, H. Thermal and bonding properties of epoxy asphalt bond coats. *J. Therm. Anal. Calorim.* **2022**, *147*, 2013–2025. [CrossRef]
75. Chen, Q.; Wang, S.; Wang, C.; Wang, F.; Fu, H.; Yang, X. Modified waterborne epoxy as a cold pavement binder:preparation and long-term working properties. *J. Mater. Civ. Eng.* **2021**, *33*, 04021079. [CrossRef]
76. Chen, Q.; Wang, C.H.; Fu, H.; Fan, Z.; Liu, L. Prediction and extreme value optimization of tensile strength of waterborne epoxy resin for road. *Mater. Rep.* **2021**, *35*, 16172–16177.
77. Zhou, Y.Y.; Ma, G.Z.; Wang, H.D.; Li, G.; Chen, S.; Wang, H.; Liu, M. Research status of pores in thermally sprayed coatings and their effects on coating performance. *Mater. Rep.* **2016**, *30*, 90–96+119.
78. Liu, L.H. Development and Application of a New Waterproof Adhesive Layer Material for Heavy Traffic Concrete Bridge Deck. Master's Thesis, Beijing University of Civil Engineering and Architecture, Beijing, China, 2018.
79. Dai, L. Effect of Temperature and Roughness on the Mechanical Properties of Concrete Bridge Deck Pavement. Master's Thesis, Chongqing Jiaotong University, Chongqing, China, 2017.
80. Chen, Z.C. Study on Pavement's Waterproofness of Concrete Bridge. Master's Thesis, Chang'an University, Xi'an, China, 2011.
81. Chen, B.; Yang, X.P.; Han, G.X. Study on the influence of bridge deck roughness on the performance of waterproof bonding layer. *Highway* **2021**, *66*, 199–204.
82. Wang, J.J.; Zhang, H.C.; Zhu, L.G. Shotblasting treatment technology for concrete bridge deck pavement. *J. China Foreign Highw.* **2011**, *31*, 170–173.
83. Wang, D.X. Research on Waterproof and Cohesive Layer of Concrete Bridge Deck Asphalt Pavement. Master's Thesis, Chongqing Jiaotong University, Chongqing, China, 2012.

84. Hu, H.B.; Liu, G.; Qian, Z.D.; Liu, Y.; Wu, X. Experimental study of waterproof bonding layer for concrete deck of Qingshan Changjiang River Highway Bridge in Wuhan. *Bridge Constr.* **2020**, *50* (Suppl. S1), 57–62.
85. Meng, C.C. Test study of waterproof binding layer for deck pavement of Poyang Lake Bridge. *World Bridges* **2011**, *2*, 57–60.

sustainability MDPI

Review

Research Progress and Performance Evaluation of Polyvinyl Alcohol Fiber Engineered Cementitious Composites

Long Cheng [1], Shaochang Chen [2], Feng Chen [3], Chaohui Wang [2,*] and Qian Chen [2]

1 Shaanxi Transportation Holding Group Technology Development Co., Ltd., Xi'an 710075, China; cl794216178@163.com
2 School of Highway, Chang'an University, Xi'an 710064, China; 15229059278@163.com (S.C.); 2016121160@chd.edu.cn (Q.C.)
3 Shaanxi Transportation Technology Consulting Co., Ltd., Xi'an 710068, China; cf4370043730@163.com
* Correspondence: wchh0205@chd.edu.cn

Abstract: Polyvinyl alcohol fiber engineered cementitious composites (PVA-ECC) have attracted wide attention due to their high toughness and narrow cracks. This review evaluated research results on PVA-ECC to further promote its research and application. The suitable length, diameter, and content of PVA fiber for ECC were recommended. The surface modification method of PVA fiber was determined and the dispersion evaluation method of PVA fiber was explored. An investigation into the mechanical properties of PVA-ECC was conducted, and the influence of PVA fiber content and fly ash content was evaluated. The improvement degree of PVA fiber on the durability of ECC and the strengthening mechanism was clarified. Results indicate that the recommended PVA fiber length for ECC was 12 mm, the diameter was 39 μm or 40 μm, and the content was 0.60–2.00%. The surface physical modification method was recommended for PVA fiber surface modification, and the mass equalization method was recommended for the evaluation of PVA fiber dispersion. With ordinary concrete, the tensile strain of PVA-ECC can be increased by 200–320 times, and the average improvement degree of flexural strength was 43%. Tensile strength and flexural strength of PVA-ECC increase first and then decrease with the increase in fiber content. With the increase in fly ash content, the tensile strength, compressive strength, and flexural strength of PVA-ECC decreased. PVA fiber content, fly ash content, and freeze–thaw medium are the main factors affecting the frost resistance of PVA-ECC. Compared with ordinary concrete, the average decreased degree of electric flux of PVA-ECC at 28 d was 21%, and the average decreased degree of carbonation depth of PVA-ECC at 3–28 d was 9–20%.

Keywords: engineering material; polyvinyl alcohol fiber engineered cementitious composites; performance evaluation; mechanical property; durability

check for updates

Citation: Cheng, L.; Chen, S.; Chen, F.; Wang, C.; Chen, Q. Research Progress and Performance Evaluation of Polyvinyl Alcohol Fiber Engineered Cementitious Composites. *Sustainability* **2023**, *15*, 10991. https://doi.org/10.3390/su151410991

Academic Editor: Hosam Saleh

Received: 25 June 2023
Revised: 10 July 2023
Accepted: 11 July 2023
Published: 13 July 2023

1. Introduction

Engineered cementitious composites (ECC) have excellent mechanical properties and durability. However, it is brittle and easy to crack, which seriously affects the safety and durability of the structure and reduces the service life of the structure. Engineering experts have explored a variety of solutions to improve its crack resistance and toughening ability, and more and more new materials are being used in ECC [1–3]. Among them, polyvinyl alcohol fiber engineered cementitious composites (PVA-ECC) have attracted wide attention because of their high toughness and narrow cracks.

Current research on PVA-ECC mainly focuses on exploring the influence of PVA fiber length, diameter, and other factors on the performance of PVA-ECC, and the surface modification methods and dispersion evaluation methods of PVA fiber. Moreover, good mechanical properties and durability are also the focus of research on PVA-ECC. Regarding the influence of PVA fiber length on the performance of PVA-ECC, Said et al. have shown that when the length of PVA used for ECC is 8 mm, the compressive strength is 62.98 MPa [4]. The PVA fiber diameter has a significant effect on the performance of the PVA-ECC. Zhang et al. pointed out that when the PVA fiber diameter was 39 μm, the ultimate tensile strain of PVA-ECC was the largest, reaching 2.6%, which showed significant strain hardening and multi-point cracking characteristics [5]. To clarify the best surface modification method for PVA fiber, Arain et al. treated the PVA fiber surface with oil and found that compared with unmodified PVA fiber, the chemical debonding between the modified PVA fiber and the matrix could be reduced to a certain extent after the oil treatment, with a decrease of up to 2.8 J/m^2 [6]. Drechsler et al. selected n-butyl aldehyde and other aldehydes to modify the surface of PVA fiber to control the interaction between fiber and matrix in the PVA-ECC [7]. The dispersion evaluation method of PVA fiber in ECC is not uniform. Zhang et al. established a fluorescence microscopic analysis technique to quantitatively evaluate the dispersion property of PVA fibers in a matrix [8]. Liu et al. assessed the dispersion of PVA fibers in the matrix based on fluorescence image analysis, using two indexes of PVA fiber dispersion coefficient and effective utilization rate [9]. In terms of the variation law of the mechanical properties of PVA-ECC, Li et al. studied the stress–strain relationship of PVA-ECC under different fly ash content conditions and found that when the fly ash content was 75%, the maximum tensile strain reached 6.23%, and the average crack width was 1.14 mm [10]. Ling et al. found that 1.5% PVA fiber increased the flexural strength of specimens by 114.4% and the fracture toughness index by 197.5% through flexural tests [11]. To explore the durability of PVA-ECC, Şahmaran et al. found that when the fly ash content was 55%, PVA-ECC showed excellent frost resistance after 300 freeze–thaw cycles [12]. Haque et al. pointed out that the incorporation of PVA fiber could reduce the internal defects of ECC, hinder the penetration and diffusion of CO_2, and effectively improve the carbonization resistance of the ECC [13]. Zhang et al. found that the chloride ion erosion depth of the PVA-ECC specimen was reduced compared with the reference specimen, and the maximum was 33.8% [14]. In summary, the current scholars have studied the factors affecting the performance of PVA-ECC, the surface modification and dispersion evaluation of PVA fiber, and the mechanics and durability of PVA-ECC, and have achieved some results. However, the influence of characteristic parameters such as length, diameter, and content of PVA fiber on the performance of the substrate needs to be further discussed to provide a reference for clarifying the technical indicators of PVA fiber for ECC. Moreover, the research on the surface modification method of PVA fiber is not deep enough, which is limited to indoor experimental research. And the evaluation method of PVA fiber dispersion is not yet mature, which can not effectively improve the mixing quality of the PVA-ECC mixture. In addition, there are many studies on the mechanical and durability properties of PVA-ECC, but its performance evolution needs to be further explored. Moreover, the strengthening mechanism of PVA fiber on ECC mechanics and durability has not yet formed a system, which needs to be summarized.

This review comprehensively evaluates current research results and applications of PVA-ECC. Technical indexes of PVA fibers for ECC are recommended. The surface modification method of PVA fiber is determined, and the dispersion evaluation method of PVA fiber is explored. Moreover, the variation of mechanical properties and durability of PVA-ECC is revealed. To sum up, this review provides a valuable reference for further research and application of PVA-ECC. The structure of the paper is shown in Figure 1.

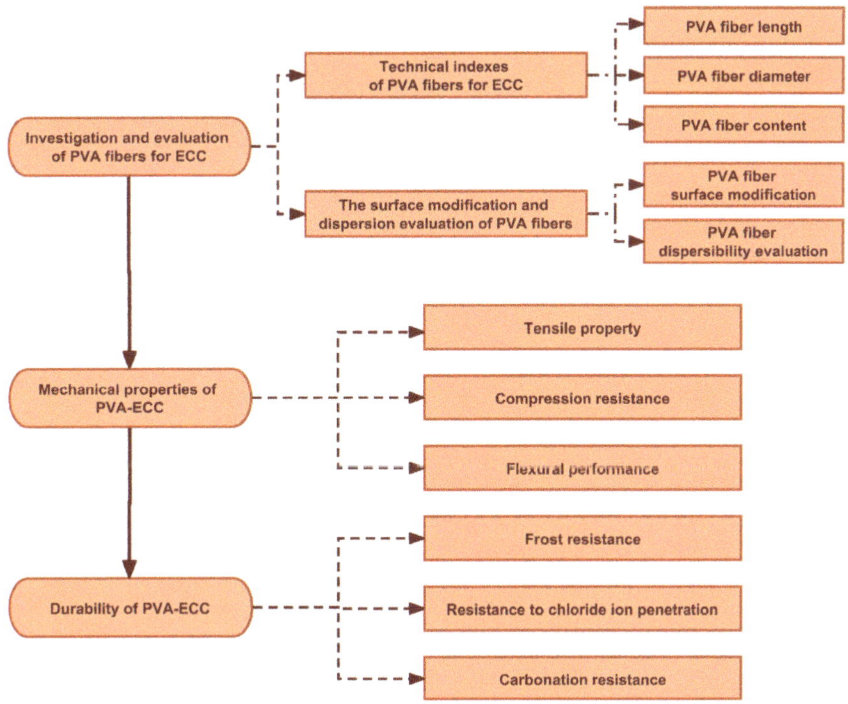

Figure 1. Structure of the paper.

2. Investigation and Evaluation of PVA Fibers for ECC

2.1. Technical Indexes of PVA Fibers for ECC

2.1.1. PVA Fiber Length

The effect of PVA fiber length on ECC is reflected in the performance. Researchers pointed out that when the PVA fiber content is the same and the length of PVA fiber is longer, the compressive strength and flexural strength of ECC will be greater [15,16]. However, Said et al. found that shorter PVA fibers could improve the compressive strength more [4]. However, Sasmal et al. considered that the influence of PVA fiber length on flexural strength was not significant [17]. It can be concluded that different experts and scholars have different research results on the mechanical properties of ECC with PVA fiber length, but most believe that longer PVA fiber is more conducive to the improvement of mechanical properties. This section discusses the PVA fiber length commonly used in studies on PVA-ECC to determine the appropriate PVA fiber length, as shown in Figure 2 [3–7,9–56].

Figure 2 shows that the PVA fiber lengths commonly used for studies on PVA-ECC mostly were 8 mm and 12 mm, accounting for 18.03% and 67.21%, respectively; by contrast, 24 mm PVA fibers accounted for only 1.64%. As can be seen, ECC in the selection of PVA fiber length, the use of 12 mm fibers is more extensive. However, the PVA fiber length should be considered and selected carefully to realize the toughening and anti-cracking effects of PVA fibers on ECC. The analysis indicates that the long length of PVA fibers incorporated into ECC is not conducive to their uniform dispersion in the ECC matrix, but rather to agglomeration, which will have a negative effect on the performance of ECC. Moreover, the more difficult and costly the process to obtain longer PVA fibers. However, when the PVA fiber is short, its bridging effect in the ECC matrix is limited, resulting in poor strengthening and toughening effect of PVA fiber on ECC. Therefore, the recommended PVA fiber length for ECC is 12 mm.

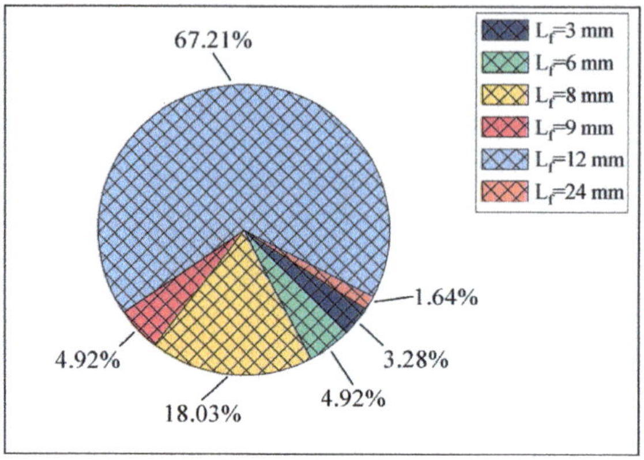

Figure 2. PVA fiber length for ECC.

2.1.2. PVA Fiber Diameter

The diameter of PVA fiber affects the bridging stress of the fibers in the ECC. It was shown that when the fiber diameter is large, PVA-ECC exhibits a significant strain hardening and multi-point cracking mode, while when the fiber diameter is small, PVA-ECC exhibits a significant strain softening and single-point cracking mode [5]. This section discusses the PVA fiber diameter commonly used in studies on PVA-ECC to determine the appropriate PVA fiber diameter, as shown in Figure 3 [3–24,26,27,29–40,42,43,45,49,50,53–57].

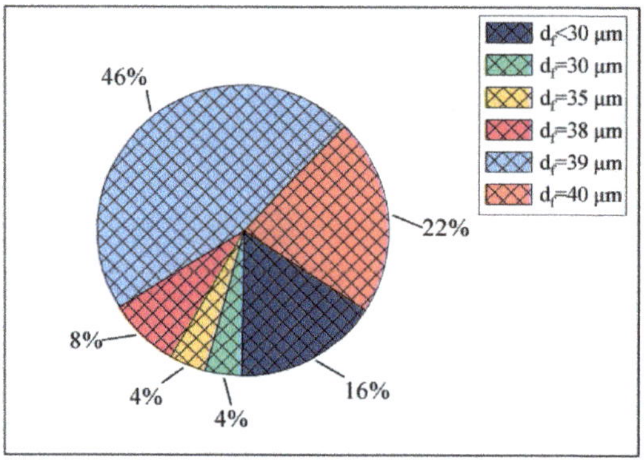

Figure 3. PVA fiber diameter for ECC.

Figure 3 shows that the diameters of PVA fibers for ECC are mostly 39 μm and 40 μm, of which 39 μm and 40 μm fibers have the highest proportion, accounting for 46% and 22%, respectively, while 35 μm and 38 μm diameter fibers account for only 4%. It can be concluded that when the diameter of PVA fiber is selected for ECC, more consideration is given to PVA fiber with a larger diameter. This is because, when the PVA fiber diameter is small, there is no guarantee that the PVA fiber is pulling out of the ECC matrix rather than breaking. However, when the diameter of PVA fiber is large, it can be achieved that most PVA fibers are pulled out rather than broken, achieving a strengthening, toughening, and crack resistance effect of PVA fibers on the ECC matrix [5]. In addition, the smaller diameter PVA fiber processing difficulties, easy to improve production costs. Therefore, it is recommended that the diameter of PVA fiber for ECC should be 39 μm or 40 μm.

2.1.3. PVA Fiber Content

Different PVA fiber contents have different effects on the performance of PVA-ECC. A critical PVA fiber content has been established. PVA fiber contents below this critical limit could not achieve the toughening and anti-cracking effect on ECC. However, when the fiber content is higher than the critical value, the PVA fibers agglomerate easily owing to the difficulty in dispersion, reducing the performance of ECC and increasing the material cost. Therefore, the PVA fiber content for ECC should be controlled within an ideal range. This section discusses the PVA fiber content commonly used in studies on ECC to determine the appropriate PVA fiber content, as shown in Figure 4 [3–6,8–17,20,21,24–45,49–56,58–61].

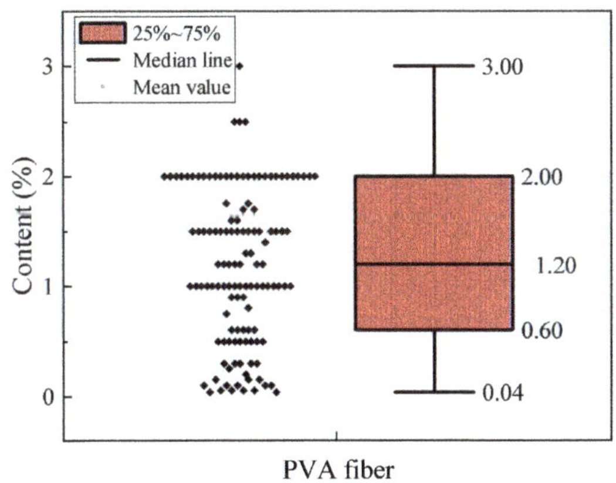

Figure 4. PVA fiber content for ECC.

Figure 4 shows a wide range of fluctuations in the PVA fiber contents used in ECC, with the highest and lowest contents at 3.00% and 0.04%, respectively. In addition, the quartile range of PVA fiber content for ECC was 0.60–2.00%, with the mean and median values at 1.20% and 2.00%, respectively. The test frequency with 1.00% content is second only to 1.20%. Accordingly, the PVA fiber content of ECC can be adopted by 2.00% for performance consideration, and 0.60% for economy consideration. Thus, the ideal PVA fiber content should be between the upper and lower quartiles of the box plot; that is, 0.60–2.00%. In addition, the above PVA fiber content is determined according to the performance requirements and economic costs of ECC. However, considering that the larger PVA fiber content is not conducive to its dispersion in the matrix. Therefore, it is suggested that the performance, cost, and dispersion of PVA fiber should be collaboratively considered in the subsequent experimental study when designing the optimum content of PVA fiber.

2.2. The Surface Modification and Dispersion Evaluation of PVA Fibers

2.2.1. PVA Fiber Surface Modification

Since the surface of PVA fiber has high hydrophilicity, a strong chemical bonding force is generated during the binding process with ECC. This results in an excessive bonding force between PVA fiber and matrix, which limits the bridging effect of PVA fiber. Furthermore, the surface modification of PVA fiber is beneficial to improve the dispersion of PVA fiber in the matrix. It is therefore important to surface modify PVA fiber to adjust and control the bonding force between PVA fiber and matrix, to improve the performance of ECC. At present, PVA fiber surface modification methods mainly focus on surface physical modification, while surface grafting modification and surface chemical modification have also made some progress, usually using environmental scanning electron microscopy (SEM), atomic force microscopy (AFM), contact angle measurement, and X-ray photoelectron spectroscopy (XPS) to evaluate the effect of PVA fiber modification.

(1) Surface physical modification

Reducing the contact between the PVA fiber and the matrix is the key to surface physical modification. Oil or emulsion forms a coating layer on the surface of PVA fiber, and changes the hydrophilic state to the hydrophobic state on the surface of PVA fiber, to adjust the adhesion between PVA fiber and matrix. This is a practical modification method. Among them, oil treatment is the most commonly used surface physical modification method at present. The oiling scheme is illustrated graphically in Figure 5. Oil treatment can reduce PVA fiber/matrix adhesion force and control fiber/matrix friction force within a certain range. However, the oil treatment has a certain limit on reducing the chemical bonding force between the PVA fiber and the matrix. In addition, research into the effects of different oil coating amounts on ECC performance is a pressing issue.

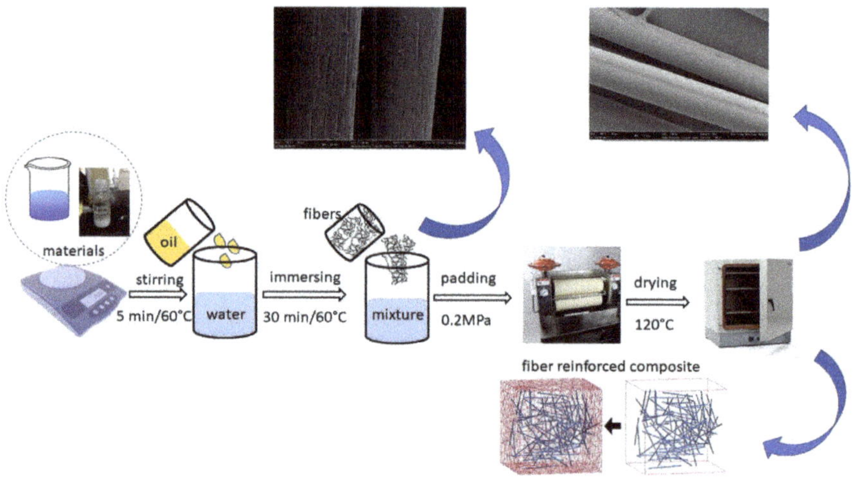

Figure 5. Fiber surface treatment illustration [19].

Silicone softener [20], styrene–butadiene emulsion [21], vinyl acetate emulsion (VAE) [58], and boric anhydride [58] are commonly used emulsions. Among them, silicone softener has good film-forming properties, which can make the original gully structure on the surface of PVA fiber evenly covered and filled under the action of silicone softener, to improve the bonding effect between PVA fiber and matrix [20]. The addition of styrene-butadiene emulsion fills the pores in the matrix and forms a thick latex film to reduce water permeability. This improves the tensile strength of PVA-ECC, up to 4.5 MPa [21]. The VAE emulsion can inhibit the hydration of the adjacent gel materials and reduce the binding force between the PVA fiber and the matrix, thus optimizing the performance of the ECC [58].

In addition to the above oil and emulsion modification, the PVA fiber can be soaked in an epoxy resin–alcohol solution first, and then the PVA fiber coated with epoxy resin is fully mixed with the nano graphite. At this time, the epoxy resin coating and the nano graphite coating on the surface of the PVA fiber can isolate the PVA fiber from the matrix, thus reducing the binding force between the PVA fiber and the matrix [22].

(2) Surface graft modification

Surface graft modification refers to connecting polymer or inorganic materials on the surface of PVA fiber by various methods so that the graft monomer can enhance the interfacial bonding effect between PVA fiber and matrix to improve the overall performance of ECC. Surface graft modification can be divided into conventional grafts and other grafts.

Conventional grafting refers to grafting polymer monomer onto the surface of PVA fiber employing the initiator. Ceric ammonium nitrate can be used as an initiator for graft modification of PVA fiber surface using glycidyl methacrylate (GMA), and the morphology of PVA fiber before and after modification is shown in Figure 6. Compared with the raw PVA fiber, it can be seen that the surface of the modified PVA fiber has a uniform scale-like structure, and the surface of the modified PVA fiber is rougher [62].

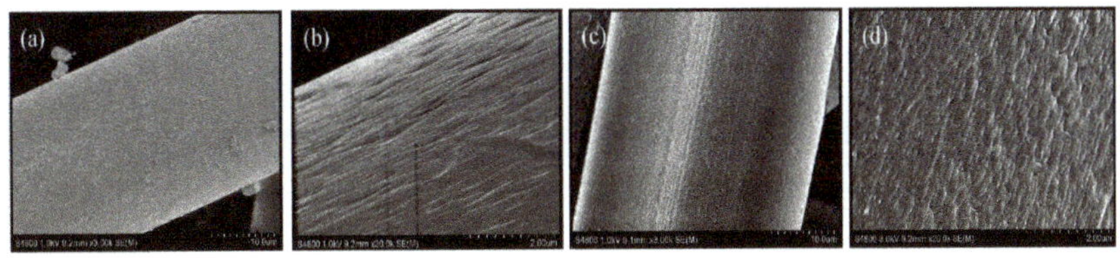

Figure 6. Morphology of PVA fiber before and after modification [62]: (**a**) raw fiber 3000 times; (**b**) raw fiber 20,000 times; (**c**) modified fiber 3000 times; (**d**) modified fiber 2000 times.

Other grafts are the grafting of inorganic substances onto the surface of the PVA fiber using a surface treatment agent or coupling agent. Polydopamine (PDA) and polyethyleneimine (PEI) can be used as surface pretreatments for PVA fibers, and finally, graphene oxide is grafted onto the surface of PVA fibers [18]. In addition, γ-(2, 3-epoxypropoxy) propytrimethoxysilane (KH560) can also be used as a coupling agent to graft nano-silica (SiO_2 NPs) onto the surface of PVA fibers. Figure 7 is the reaction mechanism schematically.

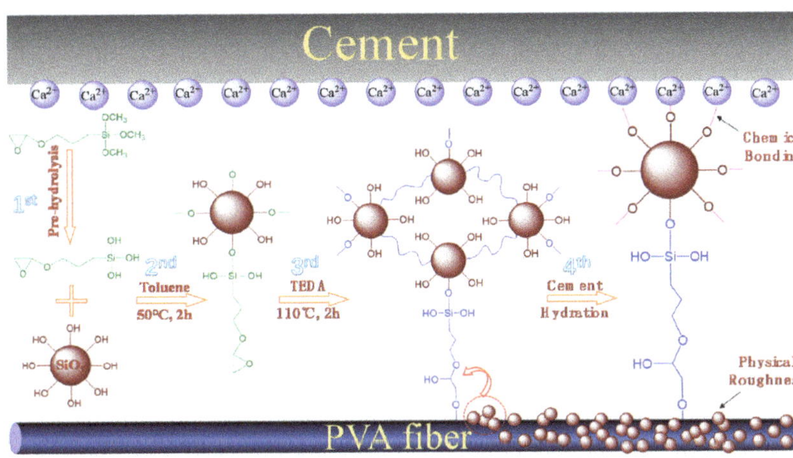

Figure 7. Schematic of the grafting procedure of SiO_2 NPs on PVA fibers and interfacial strengthening with cement upon hydration [57].

(3) Surface chemical modification

It is possible to use the hydroxyl energy on the surface of PVA fiber for acetylation reaction or by other methods and achieve the purpose of improving the bonding property of PVA fiber and matrix. Surface chemical modification mainly includes the acetalization modification method and sol–gel method.

Acetalizing modification method refers to the acetylation reaction between PVA fiber and aldehydes so that the number of the hydroxyl groups on the surface of PVA fiber is greatly reduced to achieve the purpose of regulating and controlling the bonding performance between PVA fiber and the matrix. Drechsler et al. used butanal, iso-butanal, or pentanal to modify the surface of PVA fiber [7]. Moreover, Curosu et al. found alkyl chains can be covalently bonded to the fiber surface by an acid-catalyzed acetal reaction of butyraldehyde diluted with sulfuric acid (H_2SO_4) and hydrochloric acid (HCl), as shown in Figure 8 [23].

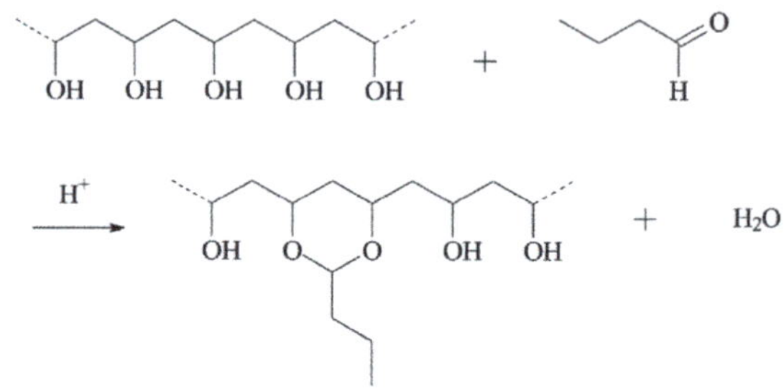

Figure 8. Reaction of butyraldehyde with PVA fiber [23].

Sol–gel method is the chemical compounds containing highly active components through solution, sol, gel, and solidification, and then after heat treatment to form oxides or other compounds solid methods. Figure 9 shows the specific steps of PVA fiber modification by sol–gel method. For example, the sol–gel method can be used to coat the surface of PVA fibers with nano-silica particles, so that the grooves on the surface of PVA fibers are fully filled with nano-silica particles, thus changing the hydrophilic state of the PVA fiber surface to a hydrophobic state and increasing the surface roughness of PVA fibers, as shown in Figure 10.

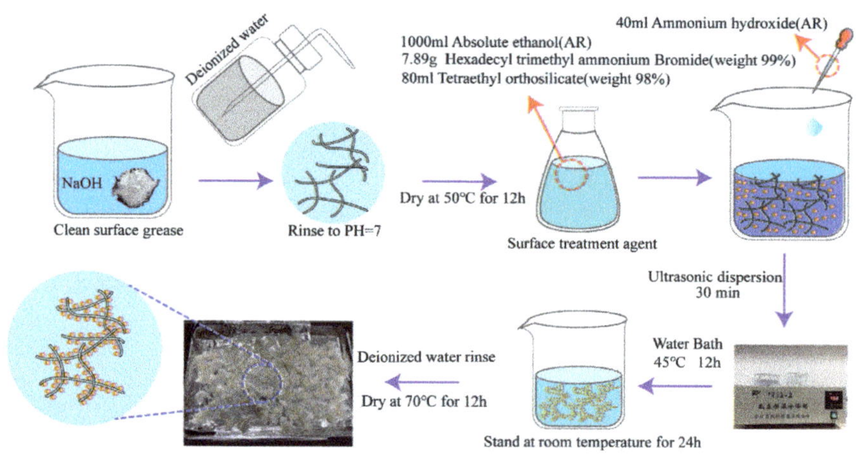

Figure 9. The schematic diagram of PVA fiber modified by the sol–gel method [24].

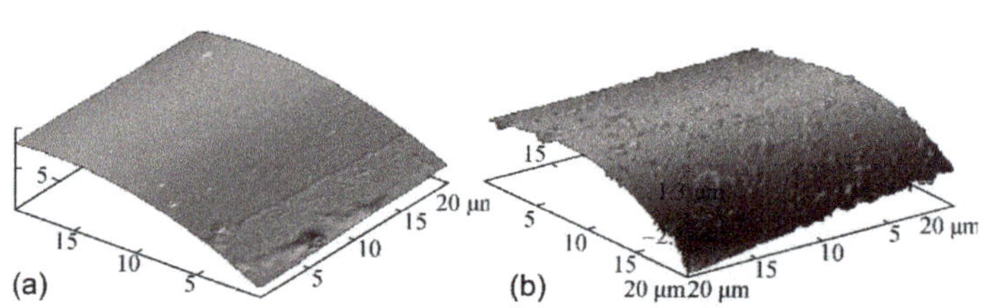

Figure 10. Surface morphology of PVA fiber [24,63]: (**a**) raw fiber; (**b**) modified fiber.

Comparing the above three PVA fiber surface modification methods, the advantages and disadvantages and improvement directions are summarized, as shown in Table 1. The three methods can be used to achieve the surface modification of PVA fiber, according to the overall analysis. However, at this time, surface modification of PVA fiber is only possible in laboratory settings, and there is a gap between low-cost and large-scale production applications. Surface physical modification is advised as a PVA fiber surface modification method after a thorough comparison of surface grafting modification, surface chemical modification, and surface modification methods.

Table 1. Comparative analysis of three surface modification methods of PVA fiber.

Modification Method	Advantage	Disadvantages	Analysis of Applicability
Surface physical modification	Simple process and has a low economic cost.	After modification, the properties of the fiber are reduced and pretreatment is needed.	The problem of fiber performance degradation after modification is further improved, which is suitable for large-scale factory processing.
Surface graft modification	Remarkable modification effect and strong flexibility.	The process is complex and the grafting rate is low.	Optimize grafting process, improve grafting rate, suitable for laboratory research and development of high-performance modified PVA fiber.
Surface chemical modification	Remarkable modification effect and cost effective.	Environmental pollution, is harmful to the human body, a long reaction time.	The selection of environmentally friendly, green, and pollution-free modified materials is currently only applicable to the laboratory scales.

The urgent area that needs to be addressed is how to further optimize the PVA fiber surface modification process under the presumption of ensuring the stability of PVA fiber performance and significant modification effect, comprehensive selection of effective and environmentally friendly modification materials, as well as clarify the influence of PVA fiber surface modification on the performance of PVA-ECC. The cooperative treatment of PVA fiber by multiple modification methods will be the future research trend because it is challenging to obtain acceptable modification effects by a single modification method.

2.2.2. PVA Fiber Dispersibility Evaluation

To prevent PVA fibers from aggregating and creating a weak interface inside the matrix, which would affect the overall performance of ECC, PVA fibers should be uniformly dispersed in the matrix. Therefore, how to qualitatively and quantitatively evaluate the dispersion of PVA fibers in ECC is an urgent problem to be solved. For the study on the dispersive evaluation method of PVA fibers, Yuan et al. used the mass equalization weighing method and gray co-incidence matrix image processing method [25]. Cao et al. used statistical methods and image processing methods [26,64]. Zhang and Liu et al. analyzed the dispersion of PVA fibers in ECC with fluorescence microscopy [6,7]. To explore the dispersion evaluation methods of PVA fiber, this section comprehensively investigated different dispersion evaluation methods and their relative applicability, as shown in Table 2 [6,7,25,26,64].

Table 2. PVA fiber dispersibility evaluation method.

Evaluation Method	Applicability
Digital image processing technique	Suitable for the interface with prominent fiber texture features and easy to recognize.
Scanning electron microscopy (SEM)	Suitable for accurately judging and grinding the fiber with a good treatment effect.
CT image gray recognition technology	Suitable for structural fiber identification.
Optical microscopy	Suitable for preliminary determination of fiber dispersion.
Mass equalization method	Suitable for quantitative evaluation of fiber dispersibility with small errors.

In conclusion, at present, the dispersion evaluation methods of PVA fiber mainly include digital image processing technology, scanning electron microscopy, CT image gray recognition technology, optical microscopy, and mass equalization method. Among them, digital image processing technology has high requirements for equipment and a complex calculation process. The effectiveness of recognition is directly related to the accuracy of evaluation results. Scanning electron microscopy (SEM) is too complicated to treat the specimen, and if the specimen is not treated properly, the imaging effect will be poor and the accuracy will be reduced. CT image gray recognition technology to identify fiber dispersion maturity is low, cannot accurately identify the identification of feature points, and the price is expensive. An optical microscope artificially observes fiber dispersion, which is more suitable for preliminary judgment. With the help of coefficient of variation calculation in mathematical statistics, the mass equalization method quantifies the dispersion effect. It has the advantages of simple equipment requirements, manual operation costs, and simple recorded data, which can reduce human errors through refined tests. The dispersion evaluation method is scientific and has a reliable inspection principle. In summary, considering the operability, accuracy, and economic cost, the mass equalization method is recommended to evaluate the dispersibility of PVA fiber in ECC.

3. Mechanical Properties of PVA-ECC

3.1. Tensile Property

Due to the incorporation of PVA fiber, ECC has superior tensile ductility than ordinary concrete material under tensile stress, showing strain-hardening and multi-crack characteristics. At present, the tests for studying the tensile properties of materials mainly include direct tensile and splitting tensile tests, and the most direct and effective test is a direct tensile test. Figure 11 summarizes the tensile strain and tensile strength from various studies to further clarify the characteristics of the tensile property of PVA-ECC and statistically regularize the results [10,27,28,30,31,59–61].

As shown in Figure 11a, the 7 d tensile strain of PVA-ECC were 1.60%, 1.99%, and 2.80% for the smaller quartile Q_1, median Q_2, and larger quartile Q_3, respectively. The interquartile range $I_{qr} = Q_3 - Q_1$ was 1.2%. The I_{qr} data indicate a low concentration degree. In addition, at 28 d, 50% of the tensile strain data of PVA-ECC ranged from 1.98–3.07%; the corresponding I_{qr} values were 1.09%. The data below the median of 2.32% were more concentrated and exhibited an overall tendency toward the Q_1 values. Referring to the ultimate tensile strain of ordinary concrete at cracking 0.01–0.02% [32,33]. It can be concluded that the 28 d tensile strain of PVA-ECC is about 200–320 times that of ordinary concrete.

Figure 11b shows that the quartile range of PVA-ECC uniaxial tensile strength at 28 d was 3.57–4.97 MPa. The I_{qr} was 1.4 MPa. The data in the 4.44–4.97 MPa range were more concentrated. Moreover, the data dispersion was large. In addition, Figure 10b also shows the quartile range of uniaxial tensile strength improvement degree of PVA-ECC at 28 d by 33–78% compared with that of ordinary concrete. The I_{qr} was 45%. The data in the 33–45% range were more concentrated. The average degree of improvement was 49%. In summary, compared with ordinary concrete, PVA-ECC has excellent tensile toughness. The analysis results suggest that the uniformly distributed PVA fiber forms a three-dimensional mesh structure in the ECC structure, which has a supporting effect on the matrix. When the ECC carries the splitting effect, the PVA fiber bears part of the tensile stress by virtue of its high strong and high modulus characteristics, which effectively alleviates the tensile stress and plays a reinforcing role. At the same time, PVA fiber gives full play to the bridging effect, prevents further cracking and damage to the ECC structure, improves the integrity and integrity of the matrix, and thus improves the ability of the ECC to resist tensile damage. In addition, cement and fly ash in the material composition of ECC improves the tensile strain capacity of the ECC structure. This is because cement and fly ash optimize the bonding performance between PVA fiber and matrix, and cement, fly ash, and PVA fiber play a synergistic role in strengthening and toughening.

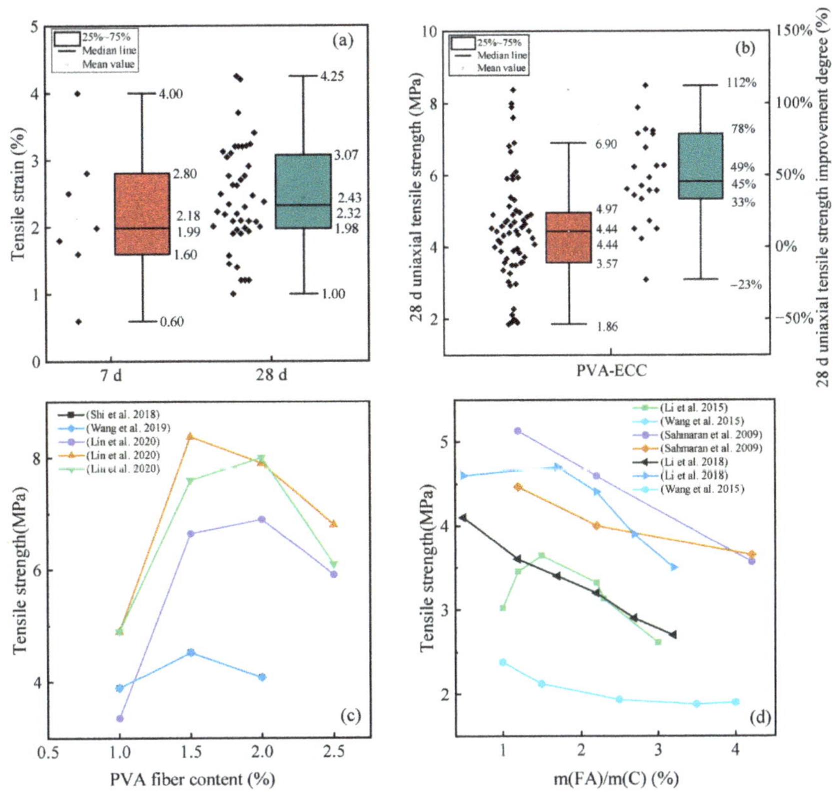

Figure 11. PVA-ECC tensile properties and influencing factors: (**a**) tensile strain; (**b**) uniaxial tensile strength; (**c**) effect of PVA fiber content on tensile strength [28,30,31]; (**d**) effect of fly ash content on tensile strength [10,59–61].

Figure 11c shows that when the content of PVA fiber is 0–2.5%, the tensile strength of PVA-ECC increases first and then decreases with the increase in PVA fiber content, and the optimal content of PVA fiber content is 1.5–2%. The analysis results suggest that when the content of PVA fiber is insufficient, the bridging effect of PVA fiber cannot be fully utilized, and its toughening and crack resistance effect is general, which cannot achieve the purpose of improving the tensile strength of ECC. However, when the PVA fiber content is too high, it is easy to cause the PVA fiber not to be uniformly dispersed in the matrix, resulting in more undesirable defects in the matrix, which makes the overall performance of ECC uneven, resulting in a decrease in the tensile strength of PVA-ECC. Therefore, further research should be carried out on the material composition of PVA-ECC. Moreover, the appropriate fiber content and how the mixing quality of the mixture can be improved should be clarified to ensure that PVA fibers positively affect the tensile strength of ECC.

Figure 11d shows that the tensile strength of PVA-ECC decreased with the increase in fly ash content. The reason is that with the increase in FA content, the cement content gradually decreases, resulting in insufficient hydration products, which weakens the strength of ECC. However, Sahmaran et al. pointed out that although the increase in FA content leads to a decrease in tensile strength, it is beneficial to obtain high tensile strain capacity, because the increase in FA content will reduce the chemical bond and matrix toughness at the fiber/matrix interface and increase the interface friction bond, thus obtaining high tensile strain capacity. In addition, the increase in fly ash content is beneficial to reduce costs. Therefore, the optimum content of fly ash should be determined according to the performance requirements of PVA-ECC and the economic cost.

3.2. Compression Resistance

Due to the absence of coarse aggregate, PVA-ECC is less stiff than ordinary concrete materials. However, the PVA fiber added plays a bridging role, which can prevent cracks from expanding inside the matrix so that extreme brittle failure will not occur, and improved the compression deformation ability of ECC. At present, PVA-ECC compressive strength is usually characterized by cylinder or cube compressive strength. Figure 12 summarizes the compressive strength from various studies to further clarify the characteristics of the compression resistance of PVA-ECC and statistically regularize the results [20,27–31,35–40,59–61,65,66].

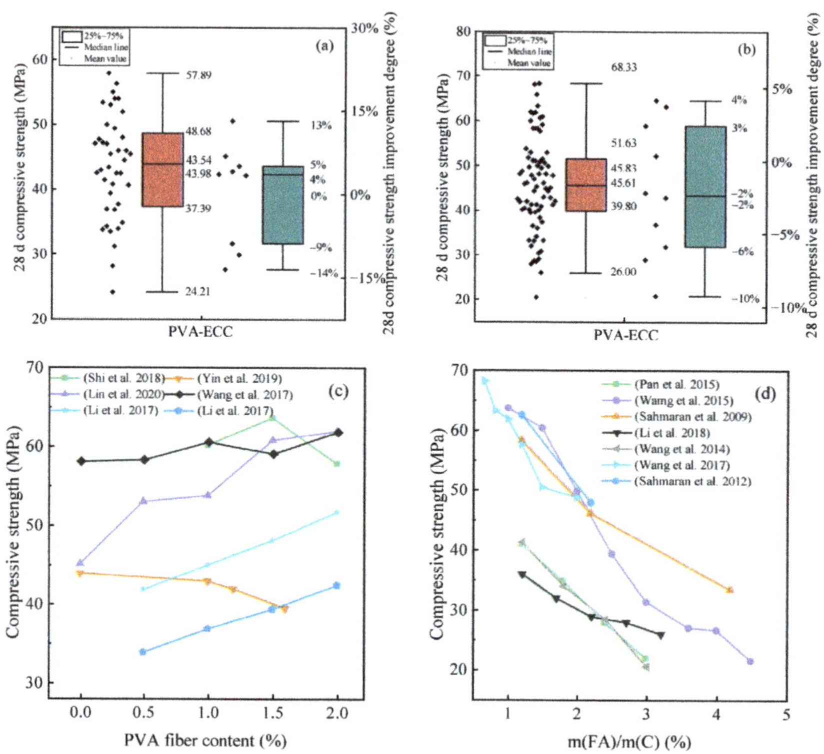

Figure 12. PVA-ECC compression properties and influencing factors: (**a**) cylinder compressive strength; (**b**) cubic compressive strength; (**c**) effect of PVA fiber content on compressive strength [28,29,31,36,39]; (**d**) effect of fly ash content on compressive strength [27,35–37,59–61].

Figure 12a shows that 50% of the data of PVA-ECC cylinder compressive strength at 28 d ranged from 37.39–48.68 MPa. The I_{qr} was 11.29 MPa. The data below the median was highly dispersed. The average value was 43.98 MPa. Figure 11b shows that the quartile range of PVA-ECC cubic compressive strength at 28 d was 39.80–51.63 MPa. The I_{qr} was 11.83 MPa. The data was symmetrical. Moreover, the dataset had a large dispersion. It can be concluded that cube specimens exhibit higher compressive strength than cylindrical specimens, which is supported by the research results of Chung et al. [40].

Figure 12a shows that 50% of the cylinder compressive strength data of PVA-ECC at 28 d improved by −9–5% compared with that of ordinary concrete. The I_{qr} was 14%. The data above the median of 4% was more concentrated and exhibited an overall tendency toward the Q_3 values. Moreover, the dataset had a large dispersion. The average degree of improvement was 0%. Figure 12b shows that the quartile range of cubic compressive strength improvement degree of PVA-ECC at 28 d by −6–3% compared with that of ordinary concrete. The I_{qr} was 9%. Moreover, the dataset had a large dispersion. The average degree of improvement was −2%. It can be concluded that compared with ordinary concrete, PVA-ECC has no obvious advantage in compressive strength, and the compressive strength of the two is close. The analysis results suggest that the PVA fibers have high hydrophilicity owing to the large number of hydroxyl groups on the surface. Therefore, in the process of bonding with ECC, PVA fiber and cement hydration products (e.g., C-S-H) are effectively bonded. This ensures the significant bridging effect of the PVA fibers. Therefore, when ECC bears compressive stress, PVA fiber gives full play to the bridging effect, and the supporting system formed by it can bear part of the load. PVA fiber plays a role in synergistic force, stress transfer, and deformation mitigation in the process of bearing deformation, thus improving the compressive strength of ECC. In addition, PVA fiber acts like an air entraining agent, introducing microbubbles into the matrix, regulating the bonding force between PVA fiber and matrix, and making PVA fiber play an active role in resisting compressive strength. However, when the content of PVA fiber is large, it is difficult to disperse the PVA fiber uniformly when the mixture is mixed, which makes the PVA fiber agglomerate in the matrix, and the coating of the cementing material gradually weakens, resulting in poor defects in the matrix, which leads to the little effect of PVA fiber on the compressive strength, and even has a negative impact on the compressive strength.

As shown in Figure 12c, at present, the research rules of PVA-ECC compressive strength with PVA fiber content are not uniform. Some studies believe that the compressive strength increases with the increase in PVA fiber content, while some studies point out that the compressive strength decreases with the increase in PVA fiber content. In fact, the two are not contradictory. This is because there are differences in the material composition design of PVA-ECC in different experimental studies. The effect of PVA fiber content on compressive strength has two opposite aspects. The positive effect is that PVA fiber plays a bridging role to inhibit the speed and width of crack expansion, so as to improve the compressive strength of ECC. The negative effect is that with the increase in PVA fiber content, PVA fiber is difficult to disperse uniformly, the pores in the matrix increase, and the density of the ECC structure deteriorates, thus reducing the compressive strength of ECC.

Figure 12d shows that the compressive strength of PVA-ECC decreases with the increase in fly ash content. The reason is that after FA replaces part of the cement, the proportion of cement decreases, resulting in fewer hydration products, and the amount of unhydrated FA increases, resulting in lower compactness of the overall structure and lower compressive strength. Therefore, in the design of fly ash content, a higher fly ash content can be selected for economic considerations, while a lower fly ash content should be considered for performance effects.

3.3. Flexural Performance

Under the action of PVA fiber bridging, the crack propagation and crack width of ECC are limited, which makes the tensile part exhibit the strain hardening characteristic of multi-crack cracking; thus, the deformation ability and bending toughness of ECC are improved [42,67]. Figure 13 summarizes the flexural strength from various studies to further clarify the characteristics of the flexural performance of PVA-ECC and statistically regularize the results [17,27–29,35–38,59,60,67,68].

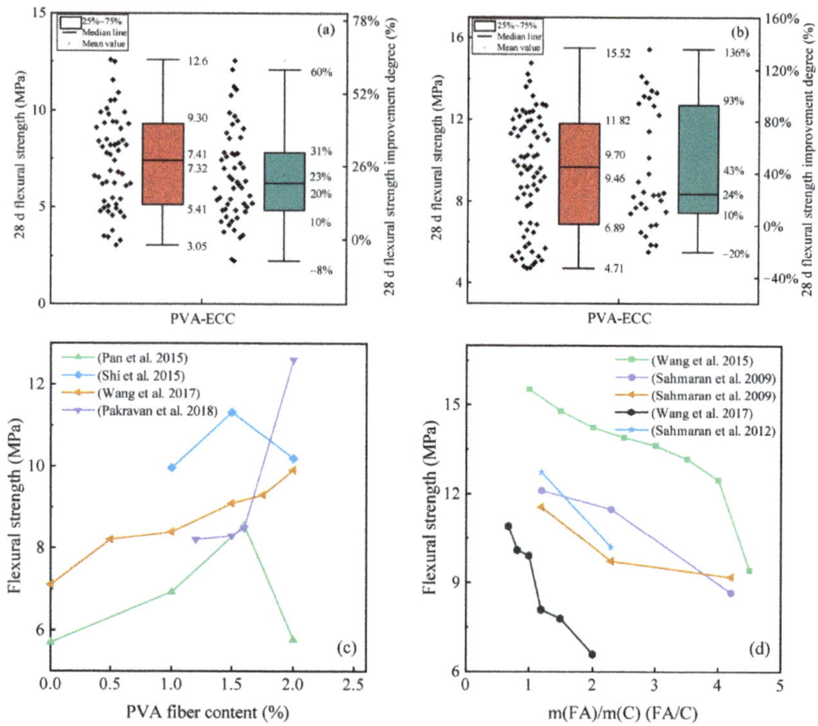

Figure 13. PVA-ECC flexural performance and influencing factors: (**a**) three-point flexural test; (**b**) four-point flexural test; (**c**) effect of PVA fiber content on flexural strength [27,28,36,67]; (**d**) effect of fly ash content on flexural strength [36,37,59,60].

Figure 13a shows that the flexural strength Q_1–Q_3 of PVA-ECC at 28 d within the three-point flexural test were 5.41 MPa, 7.41 MPa, and 9.30 MPa, respectively. The I_{qr} was 3.89 MPa. The data below the median was highly dispersed. The average value was 7.32 MPa. Figure 13b shows that the quartile range of PVA-ECC flexural strength at 28 d within the four-point flexural test was 6.89–11.82 MPa. The I_{qr} was 4.93 MPa. The data above the median 9.70 MPa was more concentrated and exhibited an overall tendency toward the Q_3 values.

Figure 13a shows the quartile range of flexural strength improvement degree of PVA-ECC at 28 d within the three-point flexural test by 10–31% compared with that of ordinary concrete. The I_{qr} was 21%. The data below the median of 20% were more concentrated. The average degree of improvement was 23%. Figure 13b shows that the flexural strength of PVA-ECC at 28 d within the four-point flexural test improved by 10–93% compared with that of ordinary concrete. The I_{qr} was 83%. The data above the median were highly dispersed. The average degree of improvement was 43%.

In conclusion, compared with ordinary concrete, PVA-ECC has higher flexural strength. This is because the PVA fiber uniformly distributed in the ECC plays a bridging role and uses its own high strength and high modulus characteristics, which has a positive impact on the flexural strength of the ECC. Under the action of bending load, the random distribution of PVA fiber effectively weakens the stress inside the matrix, and the PVA fiber across the crack gives full play to the bridging effect, inhibits the development of cracks inside the matrix, and limits the width of cracks, so that the ECC shows the strain-hardening characteristics of multi-crack cracking, and even the ECC is destroyed, though it still has good integrity, which is not similar to the brittle failure of concrete materials. Therefore, PVA-ECC has strong deformation ability and bending toughness under bending load.

Figure 13c shows that when the content of PVA fiber is 0–2.5%, the flexural strength of PVA-ECC increases first and then decreases with the increase in PVA fiber content, and the optimal content of PVA fiber content is 1.5–2%. It is concluded that when the content of PVA fiber is between 1.5–2.0%, the PVA fibers are uniformly distributed in the matrix, which not only ensures its isotropic mechanical properties, but also enables it to give full play to the bridging effect, thus playing a positive role in the flexural performance of ECC. However, when the PVA fiber content is greater than 2.0%, and the mixing quality of the mixture is poor, the PVA fiber is agglomerated, so that the PVA fiber cannot be uniformly dispersed in the matrix, and the stress concentration problem occurs inside the matrix, which has a negative effect on the flexural performance of ECC, resulting in a decrease in the ductility of ECC. Therefore, further research should be carried out on the material composition of PVA-ECC to clarify the optimum content of PVA fibers to reduce the weakening of the flexural strength due to the admixture of PVA fiber.

Figure 13d shows that the flexural strength of PVA-ECC decreased with the increase in fly ash content. The analysis shows that when the fly ash is too high, the bonding force between the PVA fiber and the matrix is too low, which limits the bridging effect of PVA fiber and leads to the decrease in flexural strength of PVA-ECC. However, the high content of fly ash increases the interfacial friction bond and improves the deformation ability of PVA-ECC, which is manifested in that the cracking of PVA-ECC under load results in its larger deformation ability and higher toughness.

4. Durability of PVA-ECC

4.1. Frost Resistance

The incorporation of PVA fibers can enhance the freeze-resistance of ECC because PVA fibers uniformly dispersed in the matrix can withstand the tensile stress caused by freezing expansion, inhibit crack propagation, and effectively resist the expansion pressure generated in the freeze–thaw cycle. In addition, PVA fiber has a similar effect to air entraining agent, and PVA fiber can introduce microbubbles into the matrix, increase the air content of the matrix, and effectively alleviate the hydrostatic pressure and osmotic pressure during the low-temperature cycle, to improve the frost resistance of ECC. At present, the frost resistance of PVA-ECC is mainly evaluated by relative dynamic elastic modulus loss rate, dynamic elastic modulus loss rate, mechanical property loss rate, and maximum freeze–thaw cycles. The frost resistance of PVA-ECC is affected by many factors.

(1) PVA fiber content

The content of PVA fiber is an important factor affecting the freezing resistance of PVA-ECC. The increase in PVA fiber content is beneficial to give full play to the fiber bridging effect to limit the crack propagation, and to improve the frost resistance of ECC. However, when the fiber content is too high, the fiber is difficult to disperse uniformly, and fiber agglomeration is easy to occur, which reduces the frost resistance of ECC. Zhao et al. considered that the optimum content was 0.15%, the mechanical properties of PVA-ECC after freeze–thaw cycles were the best, and the compressive strength loss rate was only 35% [46]. Ge et al. pointed out that the optimal content was 2% when the mass loss rate and relative longitudinal dynamic elastic modulus loss rate of PVA-ECC were the minima [47]. In view of this, it is suggested that the cost, performance, and dispersion of the fiber in the matrix should be considered comprehensively when designing the content of PVA fiber.

(2) Fly ash content

Whether the content of fly ash is moderate determines whether the addition of fly ash can improve ECC frost resistance. After the fly ash is mixed into the ECC, the fly ash particles can fill the gap between the cement particles, and the hydrated calcium silicate gel formed by the hydration of the fly ash fills the pore gap of the matrix, which improves the compactness of the ECC. At the same time, part of the hollow glass sphere of fly ash is introduced into the matrix, which will cut off the channel of capillary water seepage, thereby improving the frost resistance of ECC. However, when the content of fly ash is too high, the reaction time of volcanic ash is prolonged, and a large amount of fly ash is not hydrated, which leads to the decrease in air content in the matrix and the decrease in frost resistance of ECC. Therefore, further research should be carried out on the composition of PVA-ECC materials to determine the optimal fly ash content, and to ensure that fly ash has a positive effect on the frost resistance of ECC.

(3) Freeze–thaw medium

In addition, Du et al. pointed out that compared with water freeze–thaw, PVA-ECC had worse freeze resistance in a chlorine environment [48]. However, the effect of freeze–thaw medium on the frost resistance of PVA-ECC has not been studied in depth, and there are only experimental conclusions on macroscopic mechanical properties. It is an urgent problem to further explore the influence mechanism of different freeze–thaw media on the performance of PVA-ECC based on microscopic means.

4.2. Resistance to Chloride Ion Penetration

At present, the indicators for evaluating chloride ion penetration are not uniform, mainly including electric flux, chloride ion diffusion coefficient, water penetration height, etc. In this paper, the investigation is conducted by referring to the Standard for Test Methods of Long-term Performance and Durability of Ordinary Concrete (GB/T 50082-2009) [69] and the evaluation indexes of electric flux suggested by ASTM C1202 [70]. The evaluation standards of chlorine ion permeability of ASTM C1202 are shown in Table 3. In this section, to clarify the resistance of PVA-ECC to chloride ion penetration, the electric flux test results of PVA-ECC are summarized in Figure 14 [14,49–53,68].

Table 3. Evaluation standard of chloride ion permeability based on electric flux.

Electric Flux/C	**>4000**	**2000–4000**	**1000–2000**	**100–1000**	**<100**
Chloride ion permeability	High	Medium	Low	Lower	Can be ignored

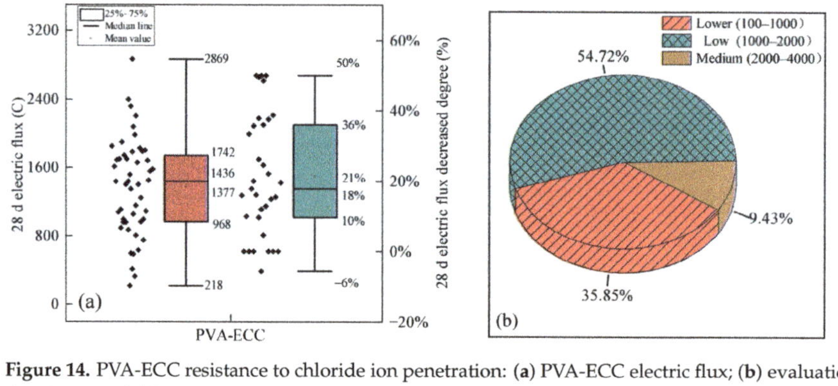

Figure 14. PVA-ECC resistance to chloride ion penetration: (**a**) PVA-ECC electric flux; (**b**) evaluation distribution of chloride ion permeability of PVA-ECC.

Figure 14a shows that the electric fluxes Q_1–Q_3 of PVA-ECC at 28 d were 968 C, 1436 C, and 1742 C, respectively. The I_{qr} was 774 C. The data in the 1436–1742 C range were more concentrated. The data dispersion was large. The average value was 1377 C. Moreover, the quartile range of electric flux decreased and the degree of PVA-ECC at 28 d is 10–36% compared with that of ordinary concrete. The Iqr values were 26%. The data dispersion was large. The average degree of decrease in the electric flux at 28 d was 21%. Figure 14b shows that in the evaluation distribution of chlorine ion permeability of PVA-ECC, 54.72% belonged to low grade, and only 9.43% belonged to medium grade.

In summary, compared with ordinary concrete, PVA-ECC exhibits excellent resistance to chloride ion penetration. The analysis results suggest that the uniform distribution of PVA fibers stably plays a bridging role and forms a supporting system inside the matrix. PVA fibers play a role of "reinforcement stabilization" and "anchoring and interlocking", effectively bridging micro-cracks and controlling crack width, improving the integrity and consistency of the matrix, reducing the chloride ion penetration and migration channels inside the ECC structure, resulting in a decrease in the amount of chloride ion penetration and migration, thereby improving the impermeability of ECC. In addition, fly ash can also improve the resistance to chloride ion erosion of PVA-ECC. Because of the filling effect of fly ash, fly ash particles fully fill the matrix voids and the internal pore size of the matrix is refined, which increases the compactness of ECC. On the other hand, the volcanic ash reaction of fly ash can produce reaction products such as hydrated calcium silicate and hydrated calcium aluminate. These products can improve the internal structure of the matrix and reduce the porosity of the matrix, thus hindering the penetration and migration of chloride ions.

4.3. Carbonation Resistance

The working environment of ECC is complex. During the service period, they are usually exposed to air. The CO_2 in the air will cause the carbonization of the material, which will reduce the durability of the structure [54]. In this section, to clarify the carbonization resistance of PVA-ECC, the carbonization test results of PVA-ECC are summarized in Figure 15 [54–56,71–73].

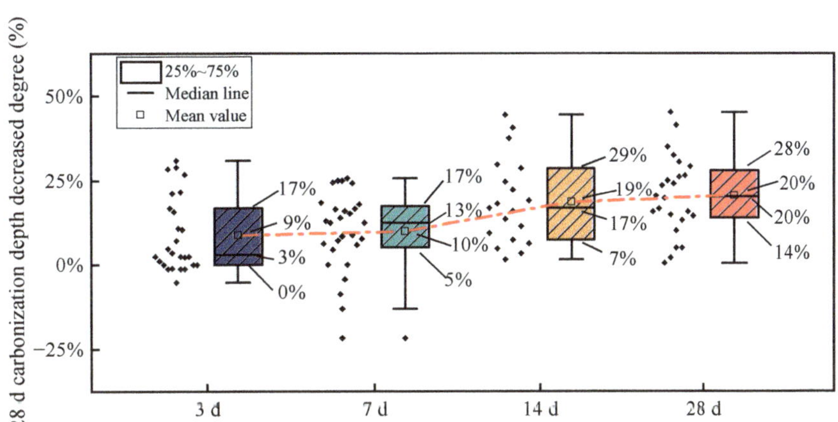

Figure 15. Carbonation resistance of PVA-ECC.

Figure 15 shows that the quartile range of carbonation depth decreased degree of PVA-ECC at 3 d, 7 d, 14 d, and 28 d is 0–17%, 5–17%, 7–29%, and 14–28% compared with that of ordinary concrete, respectively. The I_{qr} values were 17%, 12%, 22%, and 14%, respectively. The data dispersion was large. It can be concluded that compared with ordinary concrete, PVA fiber has a significant positive effect on the carbonation resistance of ECC. The analysis results suggest that the uniformly dispersed PVA fibers form a dense mesh structure in the matrix, which reduces the micropores inside the matrix to a certain extent, improves the compactness of the matrix, optimizes the internal structure of the matrix, and then reduces the diffusion channel of CO_2. On the other hand, PVA fibers efficiently play a bridging role, which inhibits the speed and width of crack expansion inside the matrix, improves the ECC structure, hinders the diffusion of CO_2, and improves the carbonation resistance of ECC. However, there is a critical value for PVA fiber to improve the carbonation resistance of ECC, which is the optimal content of PVA fiber. Below the optimum content, the bridging effect of PVA fiber is general, resulting in the strengthening, toughening, and crack resistance effect of PVA fibers on ECC not being achieved. However, higher than the optimal content, PVA fibers are difficult to disperse uniformly, resulting in more fine pores in the matrix, providing channels for the diffusion of CO_2, which has a negative effect on the carbonation resistance of ECC.

In addition, Figure 15 shows that the average degrees of decrease in the carbonation depth at 3 d, 7 d, 14 d, and 28 d were 9%, 13%, 19%, and 20%, respectively. It can be concluded that with the extension of carbonization time, the decreased degree of PVA-ECC carbonization depth increases gradually. When the carbonization time is longer (14 d, 28 d), the carbonization resistance of PVA fiber to ECC is more significant. The reason is the positive effect of the incorporation of fly ash into ECC. In the middle and late carbonization periods, the secondary reaction of fly ash gradually plays a role. The hydration products of fly ash, calcium silicate hydrate, and calcium aluminate hydrate can improve the internal structure of the matrix and make the matrix denser, thus hindering the diffusion of CO_2 and reducing the carbonization rate of ECC.

5. Conclusions

This review comprehensively discussed the application of PVA fibers in ECC. The literature review is summarized as follows:

(1) The suitable length, diameter, and content of PVA fiber for ECC were specified. It is recommended that the length of PVA fiber is 12 mm, the diameter is 39 μm or 40 μm, and the content is 0.60–2.00%.

(2) Comprehensive comparison of surface physical modification, surface grafting modification, and surface chemical modification methods, surface physical modification is recommended as a PVA fiber surface modification method. Compared with digital image processing technology, scanning electron microscopy, CT image gray recognition technology, and optical microscopy, the mass equipartition method is recommended as the dispersibility evaluation method of PVA fiber.

(3) Compared with ordinary concrete, the tensile strain of PVA-ECC can be increased by 200–320 times, the compressive strength has no obvious advantage, and the average improvement degree of flexural strength (four-point flexural) was 43%. The tensile strength and flexural strength of PVA-ECC increase first and then decrease with the increase in fiber content, the compressive strength has no obvious rule, and the optimum fiber content was recommended to be 1.5–2%. The tensile strength, compressive strength, and flexural strength of PVA-ECC decrease with the increase in fly ash content. When designing fly ash content, the performance and economy of PVA-ECC should be considered comprehensively.

(4) Compared with ordinary concrete material, PVA-ECC has excellent durability performance. PVA fiber content, fly ash content, and freeze–thaw medium were the main factors affecting the frost resistance of PVA-ECC. Compared with ordinary concrete, the average decreased degree of electric flux of PVA-ECC was 21% (28 d), and the average decreased degree of carbonation depth of PVA-ECC was 9–20% (3–28 d).

(5) The current research on the durability of PVA-ECC is weak, and the durability issues of PVA-ECC should be studied in depth, especially the anti-alkali–silica reaction. And the research on the performance of PVA-ECC mainly focuses on the macro level; the micro research on the influence of PVA fiber on the performance of ECC should be carried out in the future. Moreover, the strengthening mechanism of PVA fiber on ECC was explored using macro and micro means. In addition, the current PVA fiber has limitations in surface modification, dispersion evaluation, and performance improvement, which is not conducive to the performance improvement of PVA-ECC and the large-scale application of physical engineering. Therefore, the follow-up research on hybrid fiber engineered cementitious composites should be carried out to explore new combinations (such as PVA fiber + steel fiber, PVA fiber + PP fiber, etc.), and to clarify the performance evolution of hybrid fibers on ECC.

Author Contributions: Conceptualization, L.C., C.W. and Q.C.; methodology, L.C., C.W. and S.C.; investigation, S.C., F.C. and Q.C.; data analysis, L.C., S.C. and F.C.; visualization, S.C., F.C. and Q.C.; writing—original draft, L.C. and S.C.; review, L.C., C.W. and Q.C.; editing, L.C., S.C., C.W., F.C. and Q.C.; supervision, L.C. and C.W.; funding acquisition, L.C. and C.W. All authors have read and agreed to the published version of the manuscript.

Funding: This research was funded by the Science and Technology Project of Shaanxi Transportation Technology Consulting Co., Ltd., Innovation Capability Support Program of Shaanxi (No. 2022TD-07), China Postdoctoral Science Foundation (2023M730356).

Institutional Review Board Statement: Not applicable.

Informed Consent Statement: Not applicable.

Data Availability Statement: The data used to support the findings of this study are available from the corresponding author upon request.

Acknowledgments: This research work is supported by the Science and Technology Project of Shaanxi Transportation Technology Consulting Co., Ltd., Innovation Capability Support Program of Shaanxi (No. 2022TD-07), China Postdoctoral Science Foundation (2023M730356). That sponsorship and interest are gratefully acknowledged.

Conflicts of Interest: The authors declare no conflict of interest.

References

1. Editorial Department of China Journal of Highway and Transport. Review on China's pavement engineering research·2020. *China J. Highw. Transp.* **2020**, *33*, 1–66.
2. Wang, C.; Li, Y.; Wen, P.; Zeng, W.; Wang, X. A comprehensive review on mechanical properties of green controlled low strength materials. *Constr. Build. Mater.* **2022**, *363*, 129611. [CrossRef]
3. Liu, L.; Wang, C.; Liang, Q.; Chen, F.; Zhou, X. A state-of-the-art review of rubber modified cement-based materials: Cement stabilized base. *J. Clean. Prod.* **2023**, *392*, 136270. [CrossRef]
4. Said, S.H.; Razak, H.A.; Othman, I. Flexural behavior of engineered cementitious composite (ECC) slabs with polyvinyl alcohol fibers. *Constr. Build. Mater.* **2015**, *75*, 176–188. [CrossRef]
5. Zhang, J.; Ju, X.; Guo, Z. Tensile Properties of Fiber Reinforced Cement Composite with Different PVA Fibers. *J. Build. Mater.* **2009**, *12*, 706–710.
6. Arain, M.F.; Wang, M.; Chen, J.; Zhang, H. Study on PVA fiber surface modification for strain-hardening cementitious composites (PVA-SHCC). *Constr. Build. Mater.* **2019**, *197*, 107–116. [CrossRef]
7. Drechsler, A.; Frenzel, R.; Caspari, A.; Michel, S.; Holzschuh, M.; Synytska, A.; Curosu, I.; Liebscher, M.; Mechtcherine, V. Surface modification of poly (vinyl alcohol) fibers to control the fiber-matrix interaction in composites. *Colloid Polym. Sci.* **2019**, *297*, 1079–1093. [CrossRef]
8. Zhang, L.; Guo, L.; Sun, W.; Zhang, W.; Chen, Z. Rheological property and fiber dispersion of high ductility cementitious composites. *J. Southeast Univ. (Nat. Sci. Ed.)* **2014**, *44*, 1037–1040.
9. Liu, J.; Zhang, L.; Li, C.; Liu, J. Dispersive Characterization and Control of Fiber in Polyvinyl Alcohol Fiber Cement Composites. *J. Chin. Ceram. Soc.* **2015**, *43*, 1061–1066.
10. Li, W.; Zhou, X.; Li, N. Research on the effect of fly ash content on the tensile properties of pva-ecc. In Proceedings of the 2015 Asia-Pacific Energy Equipment Engineering Research Conference, Zhuhai, China, 13–14 June 2015.
11. Ling, Y.; Zhang, P.; Wang, J.; Chen, Y. Effect of PVA fiber on mechanical properties of cementitious composite with and without nano-SiO_2. *Constr. Build. Mater.* **2019**, *229*, 117068. [CrossRef]
12. Şahmaran, M.; Özbay, E.; Yücel, H.E.; Lachemi, M.; Li, V.C. Frost resistance and microstructure of Engineered Cementitious Composites: Influence of fly ash and micro poly-vinyl-alcohol fiber. *Cem. Concr. Compos.* **2012**, *34*, 156–165. [CrossRef]
13. Haque, A.M.; Chen, B.; Ahmad, M.R. Evaluating the physical and strength properties of fiber reinforced magnesium phosphate cement mortar considering mass loss. *Constr. Build. Mater.* **2019**, *217*, 427–440. [CrossRef]
14. Zhang, Q.; Xie, Y.; Gu, R.; Liang, X.; Zhang, Z. Chloride ion resistance of multi-scale fiber reinforced cementitious composites. *J. Civ. Environ. Eng.* **2023**, 1–10.
15. Tan, M.; Sun, R.; Zhou, Z.; Zhu, X.; Zhao, Z. Experimental study on the influence of PVA fiber length and content on the properties of engineering cement-based composites. *Railw. Eng.* **2014**, *3*, 115–117.
16. Yu, B.; Zhou, J.; Kong, Y.; Yang, W.; Cheng, B. Effect of PVA Fiber Length on Mechanical Properties of Ultra-High Toughness Cementitious Composites. *Bull. Chin. Ceram. Soc.* **2020**, *39*, 3425–3431.
17. Sasmal, S.; Avinash, G. Investigations on mechanical performance of cementitious composites micro-engineered with poly vinyl alcohol fibers. *Constr. Build. Mater.* **2016**, *128*, 136–147. [CrossRef]
18. Yao, X.; Shamsaei, E.; Chen, S.; Zhang, Q.; de Souza, F.B.; Sagoe-Crentsil, K.; Duan, W. Graphene oxide-coated Poly (vinyl alcohol) fibers for enhanced fiber-reinforced cementitious composites. *Compos. Part B Eng.* **2019**, *174*, 107010. [CrossRef]

19. Arain, M.F.; Wang, M.; Chen, J.; Zhang, H. Experimental and numerical study on tensile behavior of surface modified PVA fiber reinforced strain-hardening cementitious composites (PVA-SHCC). *Constr. Build. Mater.* **2019**, *217*, 403–415. [CrossRef]
20. Wang, M.; Muhammad, F.A.; Zhang, H.; Guo, Y. Effect of Srftace Mdifcation of Domestic Polyvinyl Alcohol Fiber on Fexural Propeties of Engineered Cementtious Compoites. *Synth. Fiber China* **2019**, *48*, 37–41+45.
21. Kim, K.-W.; Yu, C.; Han, J.-W.; Park, C.-G. Strength and durability of rapid set PVA fiber reinforced LMC for pavement repair. *Polym. Polym. Compos.* **2019**, *27*, 179–188. [CrossRef]
22. Ding, C.; Guo, L.; Chen, B.; Xu, Y.; Cao, Y.; Fei, C. Micromechanics theory guidelines and method exploration for surface treatment of PVA fibers used in high-ductility cementitious composites. *Constr. Build. Mater.* **2019**, *196*, 154–165. [CrossRef]
23. Curosu, I.; Liebscher, M.; Alsous, G.; Muja, E.; Li, H.; Drechsler, A.; Frenzel, R.; Synytska, A.; Mechtcherine, V. Tailoring the crack-bridging behavior of strain-hardening cement-based composites (SHCC) by chemical surface modification of poly(vinyl alcohol) (PVA) fibers. *Cem. Concr. Compos.* **2020**, *114*, 103722. [CrossRef]
24. Zhou, Y.; Huang, J.; Yang, X.; Dong, Y.; Feng, T.; Liu, J. Enhancing the PVA fiber-matrix interface properties in ultra high performance concrete: An experimental and molecular dynamics study. *Constr. Build. Mater.* **2021**, *285*, 122862. [CrossRef]
25. Yuan, Q.; Liang, N.; Zhao, C.; Yang, Z.; Zeng, S. Effect of PVA Fibers Dispersion on Mechanical Properties of Cement Mortar and Cement Stabilized Macadam. *J. Build. Mater.* **2021**, *24*, 921–929.
26. Cao, Y.; Li, Y.; Zheng, N.; Wang, Q.; Zhou, Y. Research on Evaluation Method of Fiber Mixing Uniformity of PVA Fiber Cement Stabilized Macadam. *Highway* **2018**, *63*, 32–35.
27. Pan, Z.; Wu, C.; Liu, J.; Wang, W.; Liu, J. Study on mechanical properties of cost-effective polyvinyl alcohol engineered cementitious composites (PVA-ECC). *Constr. Build. Mater.* **2015**, *78*, 397–404. [CrossRef]
28. Shi, L.; Lin, S.T.K.; Lu, Y.; Ye, L.; Zhang, Y. Artificial neural network based mechanical and electrical property prediction of engineered cementitious composites. *Constr. Build. Mater.* **2018**, *174*, 667–674. [CrossRef]
29. Yin, Y.; Qiu, B. Experimental Study on Mechanical Properties and Early Cracking of Polyvinyl Alcohol Fiber Concrete. *Bull. Chin. Ceram. Soc.* **2019**, *38*, 454–458.
30. Wang, Y.; Sun, L.; Gao, Y.; Liu, S. Experimental Study on Shear Performance of Oblique Section Of PVA-ECC Beams. *Ind. Constr.* **2019**, *49*, 184–188+42.
31. Lin, J.-X.; Song, Y.; Xie, Z.-H.; Guo, Y.-C.; Yuan, B.; Zeng, J.-J.; Wei, X. Static and dynamic mechanical behavior of engineered cementitious composites with PP and PVA fibers. *J. Build. Eng.* **2020**, *29*, 101097. [CrossRef]
32. Ma, H.; Cai, J.; Lin, Z.; Qian, S.; Li, V.C. $CaCO_3$ whisker modified Engineered Cementitious Composite with local ingredients. *Constr. Build. Mater.* **2017**, *151*, 1–8. [CrossRef]
33. Lin, J.; Yu, J.; Li, V.C. Mechanical properties of PVA filber reinforced engineered cementitious composites after thermal treatment. *Acta Mater. Compos. Sin.* **2016**, *1*, 116–122.
34. Gao, S.; Wang, Z.; Wang, W.; Qiu, H. Effect of shrinkage-reducing admixture and expansive agent on mechanical properties and drying shrinkage of Engineered Cementitious Composite (ECC). *Constr. Build. Mater.* **2018**, *179*, 172–185. [CrossRef]
35. Wang, W.; Pan, Z.; Meng, S.; Qiao, Z. Study on Mechanical Properties of Domestic PVA Fiber Reinforced Cementitious Composite. *Ind. Constr.* **2014**, *44*, 958–964.
36. Wang, H.; Tian, W. Experimental Study of Mechanical Properties and Frost Resistance Of PVA-ECC. *Ind. Constr.* **2017**, *47*, 123–125+130.
37. Sahmaran, M.; Yücel, H.E.; Demirhan, S.; Li, V.C. Combined effect of aggregate and mineral admixtures on tensile ductility of engineered cementitious composites. *ACI Mater. J.* **2012**, *109*, 627.
38. Huang, J.; Yang, D.; Zhu, Z.; Wang, T.; Wang, T. Reseach of the long-term mechanical properties of polyiny lalcohol fber cncrete. *Concrete* **2019**, *354*, 76–80.
39. Li, Y.; Cheng, G.; Liu, Z. Analysis of Strength And Deformation Prpperties On PVA-ECC under Uniaxial Compression. *Ind. Constr.* **2017**, *47*, 122–126+158.
40. Chung, K.L.; Ghannam, M.; Zhang, C. Effect of specimen shapes on compressive strength of engineered cementitious composites (ECCs) with different values of water-to-binder ratio and PVA fiber. *Arab. J. Sci. Eng.* **2018**, *43*, 1825–1837. [CrossRef]
41. Dias, F.G.; Segadaes, A.M.; Perottoni, C.A.; Cruz, R.C.D. Assessment of the fluxing potential of igneous rocks in the traditional ceramics industry. *Ceram. Int.* **2017**, *43*, 16149–16158. [CrossRef]
42. Li, V.C. On engineered cementitious composites (ECC) a review of the material and its applications. *J. Adv. Concr. Technol.* **2003**, *1*, 215–230. [CrossRef]
43. Huang, J.; Yang, D.; Zhu, Z.; Wang, T.; Du, B.; Wang, T. Experimental study on durblity of polyvinyl alcohol fiber concrete. *Concrete* **2019**, *356*, 61–65.
44. Al Rikabi, F.T.; Sargand, S.M.; Khoury, I.; Hussein, H.H. Material Properties of Synthetic Fiber–Reinforced Concrete under Freeze-Thaw Conditions. *J. Mater. Civ. Eng.* **2018**, *30*, 04018090. [CrossRef]
45. Deng, Z.; Xuan, H.; Xu, H. Experimental Study of Durability Properties of High Toughness Fiber Reinforced Cementiitous Composites under Freezing and Thawing Cycles. *J. North China Univ. Water Resour. Electr. Power (Nat. Sci. Ed.)* **2013**, *34*, 16–19.
46. Zhao, X.; Wang, X.; Qiao, H.; Huang, R. Study on Freeze-thaw Cycle Damage Law and Model of PVA Fiber Reinforced Concrete. *China Concr. Cem. Prod.* **2022**, 53–57. [CrossRef]
47. Ge, W.; Cai, C.; Ji, X.; Ashour, A.F.; Cao, D. Experimental study of mechanical properties of PVA-ECC under freeze-thaw cycles. *J. Test. Eval.* **2018**, *46*, 2330–2338. [CrossRef]

48. Du, X.; Li, J.; Zhang, X.; Zhao, X.; Zhou, Z. Durability of Concrete with Low Content Polyvinyl Alcohol Fibers. *J. Univ. Jinan (Sci. Technol.)* **2017**, *31*, 371–376.
49. Zhong, J.; Wang, Z.; Gao, Q. Chloride ion resistance of cementitious composites with PVA fiber reinforcement. *Concrete* **2019**, *356*, 105–108.
50. Ding, X.; Zhang, J.; Wang, Q. Chloride Ion Permeability of Low Shrinkage Engineered Cementitious Composite. *J. Build. Mater.* **2017**, *20*, 827–834.
51. Yan, C.; Cao, Y.; Liu, S.; Wang, X.; Pang, P. Experimental study on salt erosion resistance of PVA fiber concrete. *Concrete* **2020**, *374*, 56–60.
52. Li, L.; Tan, Y.; Li, K.; Yang, L.; Du, X. Comparative Study on Effect of Basalt Fiber, Polyacrylonitrlie Fiber, and Polyvinyl Alcohol Fiber on Concrete Performance. *Railw. Eng.* **2019**, *59*, 130–134.
53. Si, W.; Feng, X. Experimental Study on Flexural Strength and Chloride Ion Permeability of PVA-ECC. *Bull. Chin. Ceram. Soc.* **2018**, *37*, 2214–2217.
54. Cai, X. Experimental Research on the Durable Properties of Ultra High Toughness Cementitious Composites (UHTCC). Master's Thesis, Dalian University of Technology, Dalian, China, 2010.
55. Bao, W.; Li, L.; Yuan, L. Experiment on carbonation properties of green ductile cementitious composite. *J. Shenyang Univ. Technol.* **2018**, *40*, 104–108.
56. Luo, Z. Study on Effect of PVA Fiber Modification on Carbonation Resistance of Concrete. Master's Thesis, Huazhong University of Science and Technology, Wuhan, China, 2021.
57. Zhang, W.; Zou, X.; Wei, F.; Wang, H.; Zhang, G.; Huang, Y.; Zhang, Y. Grafting SiO_2 nanoparticles on polyvinyl alcohol fibers to enhance the interfacial bonding strength with cement. *Compos. Part B Eng.* **2019**, *162*, 500–507. [CrossRef]
58. Sun, M.; Chen, Y.; Zhu, J.; Sun, T.; Shui, Z.; Ling, G.; Zhong, H.; Zheng, Y. Effect of modified polyvinyl alcohol fibers on the mechanical behavior of engineered cementitious composites. *Materials* **2018**, *12*, 37. [CrossRef]
59. Wang, H.; Tian, W.; Qing, L.; Xu, L. Dosage of Fly Ash on the Mechanical Properties of PVA Cementitious Composites. *Bull. Chin. Ceram. Soc.* **2015**, *34*, 2997–3003.
60. Sahmaran, M.; Lachemi, M.; Hossain, K.M.A.; Ranade, R. Influence of aggregate type and size on ductility and mechanical properties of engineered cementitious composites. *ACI Mater. J.* **2009**, *106*, 308.
61. Li, Y.; Zhu, J.; Wu, Y.; Yu, P. Research of Mechanics Properties of PVA Fiber Cementitions Composites. *J. Beijing Univ. Technol.* **2018**, *44*, 553–560.
62. Ke, Y.; Shen, Y.; Fei, G.; Wang, H. Surface Modification of Water Soluble Polyvinyl Alcohol Fiber with Glycidyl Methacrylate and Its Strengthening Mechanism. *Polym. Mater. Sci. Eng.* **2013**, *29*, 103–106+111.
63. Ding, C.; Guo, L.; Chen, B.; Deng, Z.; Huang, G. Alkali Resistance and Surface Modification of High Strength and High Modulus PVA Fibers. *J. Chin. Ceram. Soc.* **2019**, *47*, 228–235.
64. Cao, Y.; Zhao, Y.; Wang, Q.; Xia, Z.; Zheng, N. Mixng Uifomiy of Cement Stililed Macadam with PVA Fiber. *J. Chongqing Jiaotong Univ. (Nat. Sci.)* **2018**, *37*, 29–33.
65. Zhou, J.; Pan, J.; Leung, C.K.Y. Mechanical behavior of fiber-reinforced engineered cementitious composites in uniaxial compression. *J. Mater. Civ. Eng.* **2015**, *27*, 04014111. [CrossRef]
66. Zhang, W.; Yin, C.; Ma, F.; Huang, Z. Mechanical properties and carbonation durability of engineered cementitious composites reinforced by polypropylene and hydrophilic polyvinyl alcohol fibers. *Materials* **2018**, *11*, 1147. [CrossRef]
67. Pakravan, H.R.; Jamshidi, M.; Latifi, M. The effect of hydrophilic (polyvinyl alcohol) fiber content on the flexural behavior of engineered cementitious composites (ECC). *J. Text. Inst.* **2018**, *109*, 79–84. [CrossRef]
68. Zhong, J.F. Multi-Objective Optimization and Working State Analysis of Cement-Based Composite with PVA Fiber. Ph.D. Thesis, Harbin Institute of Technology, Harbin, China, 2020.
69. *GB/T 50082-2009*; Standard for Test Methods of Long-Term Performance and Durability of Ordinary Concrete. China Architecture and Building Press: Beijing, China, 2010.
70. *ASTM C1202-2010*; Standard Test Method for Electrical Indication of Concrete's Ability to Resist Chloride Ion Penetration. American Society for Testing Material (ASTM): West Conshohocken, PA, USA, 2010.
71. Zhang, P.; Wan, J.; Su, H.; Dai, X. Carbonation Resistance of Nano-particles and PVA Fiber Reinforced Cement Based Composites. *Water Resour. Power* **2017**, *35*, 108–111.
72. Zhang, P.; Li, Q.; Wang, J.; Shi, Y.; Ling, Y. Effect of PVA fiber on durability of cementitious composite containing nano-SiO_2. *Nanotechnol. Rev.* **2019**, *8*, 116–127. [CrossRef]
73. Luo, Z.; Yang, X.; Ji, H.; Zhang, C. Carbonation Model and Prediction of Polyvinyl Alcohol Fiber Concrete with Fiber Length and Content Effects. *Int. J. Concr. Struct. Mater.* **2022**, *16*, 9. [CrossRef]

sustainability

MDPI

Article

Properties of Concrete Reinforced with a Basalt Fiber Microwave-Absorbing Shielding Layer

Aqing Jiang [1], Zihao Song [1], Xuancang Wang [1,*], Jing Zhao [1] and Junru Ren [2]

[1] School of Highway, Chang'an University, Xi'an 710064, China; jaq20220520@163.com (A.J.); s614993780@163.com (Z.S.); zhaojingzi0203@163.com (J.Z.)

[2] Department of Military Installations, Army Logistical Academy of PLA, Chongqing 401331, China; 2021221313@chd.edu.cn

* Correspondence: wxc2005@chd.edu.cn

Abstract: The purpose of this study was to propose a highly efficient, durable, and environmentally friendly method for the rapid removal of ice and snow. A microwave-absorbing functionality layer was placed between a conductive metal mesh and magnetite sand shielding layer, and ordinary cement concrete (OC). Microwave heating, mechanical strength determination, and indoor and outdoor de-icing tests were performed on the cement concrete specimens with the shielding layer. Basalt fibers were added to the absorbing functionality layer, and the formed specimens were tested for strength and durability. The microstructure was observed using SEM experiments. The results show that the temperature rise of microwave-absorbing cement concrete with a magnetite sand shielding layer (MCMS) and microwave-absorbing cement concrete with a conductive metal mesh shielding layer (MCMM) increased by approximately 17.2% and 27.1%, respectively, compared to that of microwave-absorbing concrete (MAC). After freeze–thaw cycles, the compressive strength and flexural strength of microwave-absorbing concrete with basalt fiber (MAB) increased by 4.35% and 7.90% compared to those of MAC, respectively. The compressive strength and flexural strength of microwave-absorbing concrete with a magnetite sand shielding layer and basalt fiber (MAMB) increased by 8.07% and 6.57%, respectively, compared to those of MCMS. Compared to specimens without basalt fiber, the wear rate per unit area of MAMB decreased by 8.8%, and the wear rate of MAB decreased by 9.4%. The water absorption rate of MAMB specimens decreased by 13.1% and 12.0% under the conditions of 20 and 40 microwave freeze–thaw cycles, respectively, compared to that of MCMS. The water absorption rate of MAB specimens decreased by 9.9% and 8.3% under the conditions of 20 and 40 microwave freeze–thaw cycles, respectively, compared to that of MAC. SEM analysis showed that the addition of basalt fibers improved the compactness and stability of the cement concrete structure as a whole. This study provides valuable references for the promotion and application of microwave de-icing technology.

Keywords: microwave-absorbing concrete cement; shielding layer; basalt fiber; mechanical strength; durability

check for updates

Citation: Jiang, A.; Song, Z.; Wang, X.; Zhao, J.; Ren, J. Properties of Concrete Reinforced with a Basalt Fiber Microwave-Absorbing Shielding Layer. *Sustainability* **2023**, *15*, 15919. https://doi.org/10.3390/su152215919

Academic Editor: Constantin Chalioris

Received: 13 September 2023
Revised: 25 October 2023
Accepted: 2 November 2023
Published: 14 November 2023

1. Introduction

The formation of ice on roads threatens the safety of people, increasing the likelihood of vehicular accidents and having a detrimental effect upon society. Therefore, measures should be taken to remove ice and snow from road surfaces [1–3] to avoid traffic delays and economic losses [4,5]. Traditional de-icing methods primarily involve mechanical and chemical approaches to remove accumulated snow. However, in severe conditions, mechanical de-icing may not be feasible. These methods not only have low efficiency but can also cause chemical corrosion, thereby reducing the durability of the pavement [6]. Therefore, there is an urgent need for efficient and environmentally friendly methods to quickly remove ice and snow from road surfaces. Microwave heating, as an environmentally friendly snow melting and de-icing technology, has been applied in snow melting and de-icing [7]. Moreover, microwave technology has the unique advantage of selectively heating the road surface by penetrating the ice layer [8], offering broad prospects for rapid ice and snow removal on roads.

Micheli et al. [9,10] (2014, 2017) found that carbon nanotubes played a role in improving the electromagnetic properties of concrete, as they can shield electromagnetic waves. Wang et al. [11,12] discovered that magnetite improved the temperature increase performance of mortar materials under microwave irradiation. Liu et al. [13] used recycled cathode ray tube funnel glass as a substitute material for magnetite sand in radiation-shielding concrete. Concrete with fine aggregate replacement rates of 20–40% exhibited good performance. Guo et al. [14] designed a microwave-absorbing layer with a certain thickness of magnetite added to ordinary asphalt mixtures. The results showed that reducing the thickness of the absorbing layer did not lead to a loss in the surface temperature efficiency of the specimens. Gallego et al. [15] found that steel fibers could be used as absorbing materials to prepare absorbing asphalt concrete. After adding 10 mm long steel fibers with a mass fraction of 0.2% to asphalt concrete, the temperature reached 140 °C after 2 min of microwave heating. Lu et al. [16] incorporated carbon fiber materials into cement concrete to alter its electromagnetic characteristics. The research findings indicated that the thermal efficiency of microwave heating increased initially and then decreased with an increase in the carbon fiber content. Considering the de-icing effect and road performance of concrete, the optimal content of carbon fiber was determined to be 1‰. However, due to uneven concrete mixing, the uneven distribution of fibers in concrete significantly affected the temperature uniformity of the absorbing concrete. Liu et al. [17] investigated the effects of various factors, such as different materials and external conditions, on the efficiency of microwave de-icing. They found that the efficiency of microwave de-icing can be significantly improved by using a power of 7000 watts, a magnetic content of 60% by volume for the distributed microwave-absorbing surface, a shielding layer on the bottom surface, and wet conditions. Wang et al. [18,19] completely replaced limestone with magnetite. The results showed a significant improvement in the mechanical strength, dielectric loss, magnetic loss, and temperature increase performance of concrete. Magnetite exhibits excellent microwave-absorbing properties and has been widely applied in many studies on microwave de-icing. Zhao et al. [20] designed nucleus–shell structural microwave-enhanced aggregate (NSSA) consisting of an external microwave-absorbing and heat-collecting structure layer (ESL) to enhance the microwave absorption capacity of asphalt mixtures and improve the utilization rate of microwave energy. They also found that the NSSA architecture has better electromagnetic field distribution characteristics. Chung et al. [21] found that the shielding efficiency of transparent electromagnetic interference shielding films increased with a decrease in the aperture size and an increase in the height of the metal mesh electrode. Li et al. [22] found that a basalt fiber content of 1.5% in cement yielded good performance and improved durability. Wang et al. [23] studied the effects of three powders (Fe_3O_4, SiC, and graphite) on the efficiency of microwave heating and frost resistance. They found that all three powders can improve the efficiency of microwave heating, and SiC can improve the frost resistance of cement concrete. Fu et al. [24] explored an electromagnetic multifunctional asphalt pavement that can be heated for self-healing and snow melting through energy conversion.

Overall, research on microwave de-icing primarily focuses on improving the microwave absorption efficiency of road surfaces through the addition of coatings, selection of microwave-absorbing materials, and design of mix proportions. The goal is to effectively remove ice and snow from road surfaces. Additionally, studies on basalt fibers mainly concentrate on the dosage, fiber length, and distribution characteristics of the fibers. Research on the enhancement of strength and durability in microwave-absorbing cement concrete with basalt fibers is limited. In addition, research on improving microwave de-icing efficiency by modifying the road structure is scarce. Moreover, microwave-absorbing road surfaces undergo multiple cycles of freezing and melting in practical applications, which may result in inadequate strength and durability. Therefore, the aim of this study was to enhance microwave de-icing efficiency by introducing a microwave shielding layer beneath the microwave-absorbing functionality layer and investigating the reinforcement effect of basalt fibers on the mechanical strength and durability of concrete specimens.

Different materials were selected in order to prepare MAC specimens with microwave shielding layers. The de-icing performance of different structures was evaluated through temperature rise performance, mechanical properties, and indoor and outdoor de-icing tests. The addition of basalt fibers to the cement concrete specimens was also investigated, and tests were conducted to evaluate mechanical strength and durability. SEM analysis was employed to study the microstructure at the interface between basalt fibers and cement paste. This study provides references for the design of MAC road structures, improvement of microwave de-icing efficiency, and enhancement of structural durability. Furthermore, it promotes the application and popularization of microwave de-icing technology in road engineering.

2. Materials and Methods

2.1. Materials

2.1.1. Ordinary Materials

Silicate cement P.C.32.5 from Huaxia Cement Co., Ltd. (Huangshi, China) was used, and its performance indicators can be found in Table 1. A naphthalene-based water reducer was employed as the admixture, while limestone was used as the coarse aggregate, with performance indicators listed in Table 2. The basalt fiber utilized in this study was provided by Shanghai Chenqi Chemical Technology Co., Ltd. (Shanghai, China). The basic properties of magnetite are presented in Table 3, and the basic properties of basalt fiber can be found in Table 4. Fine sand has an apparent density of 2603 kg/m^3 and belongs to Zone II. All materials used in the study met the technical requirements specified in JTG D40—2011 [25].

Table 1. Properties of cement.

Model	Density (g/cm^3)	Specific Surface Area (m^2/kg)	Setting Time (min)		Compressive Strength (MPa)		Flexural Strength (MPa)	
			Initial Setting	Final Setting	3 d	28 d	3 d	28 d
P.C.32.5	3.15	359	128	279	12.4	37.9	3.1	6.8

Table 2. Properties of limestone.

Apparent Density (g/cm^3)	Crushing Value	Needle Flake Content (%)	Water Absorption (%)	Mud Content (%)	Firmness (%)
2.64	17	6	1.0	0.8	6

Table 3. Properties of magnetite.

Apparent Density (g/cm^3)	Bulk Content (%)	Mud Content (%)	Iron Content (%)	Mohs Hardness
3.78	2.77	0.2	65~85	6

Table 4. Properties of basalt fiber.

Length (mm)	Single Fiber Diameter (μm)	Density (g/cm^3)	Tolerance to Maximum Temperature (°C)	Elastic Modulus (G Pa)	Tensile Strength (MPa)
12	14	2.73	625	95~105.4	3250~3800

2.1.2. Shielding Layer Material

In this study, a galvanized iron wire mesh with a diameter of 0.1 cm was chosen as the conductive metal grid for microwave shielding. The mesh was formed into a grid with a side length of 1 cm. The second type of electromagnetic wave shielding layer consisted of cement concrete prepared by combining hematite and magnetite sand. An optimal thickness of 5 cm is recommended for the shielding concrete layer. Magnetite sand was obtained by crushing and grinding magnetite ore and the properties of hematite ore can be found in Table 5.

Table 5. Properties of hematite ore.

Needle Flake Particle Content (%)	Crushing Value Index	Alkali Activity (%)
2	19.5	0.06

2.2. *Specimen Preparation*

2.2.1. OC and MAC Specimens

According to the mix proportions (Table 6), standard compressive specimens with dimensions of 15 cm × 15 cm × 15 cm and standard flexural specimens with dimensions of 10 cm × 10 cm × 40 cm were prepared. The specimen preparation, curing, and testing followed the guidelines specified in JTG 3420–2020 [26].

Table 6. Material consumption per m^3 of concrete.

Specimen Types	Cement (kg)	Water (kg)	Sand (kg)	Coarse Aggregate (kg)				Magnetite (16–19 mm) (kg)	Water-Reducing Agent (kg)
				4.75–9.5 (mm)	9.5–16.0 (mm)	16.0–19.0 (mm)	19.0–26.5 (mm)		
OC	375	3.75	739	255	255	370	277	—	3.75
MAC	375	3.75	739	255	255	—	277	530	3.75

First, well-mixed ordinary cement concrete was placed in a standard mold and compacted via vibration until a 5 cm thickness was achieved. Then, the MAC mixture was prepared according to the specified proportions and placed in the standard mold on top of the previously compacted layer. The mixture was compacted again via vibration to ensure density and prepare the MAC specimens.

2.2.2. Cement Specimens with a Shielding Layer

MCMM specimens: Firstly, well-mixed ordinary cement concrete material was placed in a mold and compacted via vibration to achieve density. Then, a 3 cm thickness was reserved for the microwave absorption functional layer. The preparation of the test specimens is shown in Figure 1a,b. A galvanized iron wire mesh was placed in the mold, followed by pouring and compacting the microwave-absorbing functionality layer containing magnetite material.

MCMS specimens: Firstly, cement concrete material with a magnetite sand shielding layer (Material consumption per m^3 showed in Table 7) was prepared according to the specified mix proportions. It was then poured into the mold and compacted by vibration, reserving a 3 cm thickness. Subsequently, the microwave-absorbing functionality layer containing magnetite material was poured into the mold and compacted.

The preparation process for the flexural test standard specimens was the same as described above. After the preparation of the compressive strength test specimens and flexural strength test specimens, they were placed in a curing box for curing.

Figure 1. (**a**) compressive strength specimen; (**b**) flexural strength specimen; (**c**) basalt fiber; (**d**) ice layer preparation.

Table 7. Material consumption per m^3 of magnetite sand.

Cement (kg)	Sand (kg)	Coarse Aggregate (kg)			Hematite (kg)		Magnetite Sand (kg)	Water Reducing Agent (kg)
		4.75–9.5 (mm)	9.5–16 (mm)	19–26.5 (mm)	9.5–16 (mm)	16–19 (mm)		
375	622	255	182	197	141	715	644	3.75

2.2.3. Specimen Preparation with Basalt Fibers

When preparing cement specimens containing basalt fibers, the basalt fibers were added to the microwave-absorbing functionality layer at a volume dosage of 0.2% and mixed. The remaining steps were the same as described above.

2.2.4. Ice Layer Preparation

Firstly, approximately 1.9 cm of water was poured into a rubber mold. The cured specimens were placed in the mold, and then the mold was securely placed in a freezer. After the cooling process was completed, the mold was removed, resulting in cement specimens with an ice layer (Ice preparation as shown in Figure 1d).

2.3. Methods

Based on the conventional microwave-absorbing road structure (Figure 2a), a microwave shielding layer was introduced between the absorbing layer and the road surface layer (Figure 2b) to enhance the utilization of microwaves. The research included tests on compressive strength and de-icing effectiveness. Basalt fiber (Figure 2c) was added to the absorbing layer to improve the compressive strength and durability of the road surface. Scanning electron microscopy experiments were conducted to analyze the microscopic mechanism.

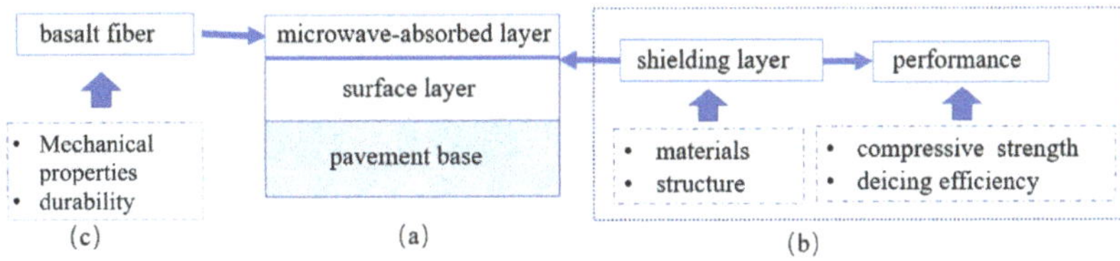

Figure 2. Design method for microwave-absorbing pavement structure. (**a**) Pavement structure, (**b**) shielding layer design, (**c**) basalt fiber reinforcement.

2.3.1. Heating Effect

Each group of OC, MAC, MCMS, and MCMM specimens consisted of three parallel specimens. The experimental results were obtained by averaging the values. After the preparation of the standard specimens, they were heated in a microwave oven (Foshan, China) with a frequency of 2.45 GHz and a maximum power of 900 W. The dimensions of the microwave oven were 42 cm × 42 cm × 25 cm (length × width × height). During the heating process, temperature measurements were taken every minute using an infrared thermometer (FLUKE-62MAX model, Tuodapu, Dongguan, China). Three measurement points were selected on the surface of the cement specimens (Figure 3a) to calculate the average temperature and heating rate of the top surface of the specimens.

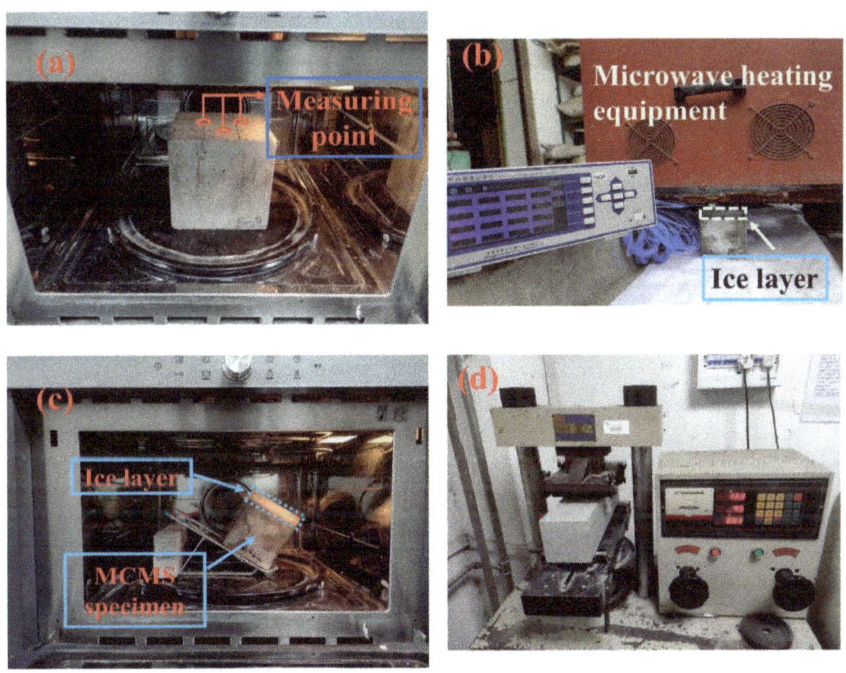

Figure 3. Experiment involving microwave-absorbing concrete with shielding layer. (**a**) Heating effect experiment; (**b**) outdoor de-icing effect test; (**c**) indoor de-icing effect test; (**d**) flexural strength experimental machine (YES-300B digital display pressure testing machine, Shaoxing Kent Mechanical Electronics Co., Ltd., Shaoxing, China).

2.3.2. De-Icing Effect

For OC, MAC, MCMS, and MCMM pavement structures, standard specimens measuring 10 cm × 10 cm × 10 cm were prepared. A 2 cm thick ice layer was formed on the surface of each specimen. Each group consisted of three parallel specimens, and the experimental results were obtained by averaging the values. Subsequently, the specimens were placed on a support structure with a certain inclined angle (Figure 3c). The microwave oven door was closed, and the time required for the ice layer to detach from the surface of the specimen during microwave heating was recorded as the indicator for evaluating the microwave de-icing efficiency.

For outdoor de-icing experiments, OC, MAC, MCMS, and MCMM specimens were prepared, with each group consisting of three standard specimens. A 2 cm thick ice layer was formed on the surface of each specimen. A portable microwave vehicle was used for heating the standard specimens (Figure 3b), with the waveguide mouth positioned 2 cm away from the concrete specimen surface. The microwave vehicle was started, and the temperature changes of the specimens during the movement of the microwave vehicle were recorded using the SH-X multi-channel temperature recorder and temperature sensing wires. The heating rate of the concrete specimen surface was used as the indicator for evaluating the microwave de-icing efficiency of the pavement structure in outdoor de-icing experiments. The time required for the top surface of the specimen to reach 0 °C was recorded using temperature sensors.

2.3.3. Mechanical Properties

OC, MAC, MCMS, and MCMM specimens were prepared separately. Three parallel specimens were prepared for each group, and the experimental results were averaged. The experiments were conducted according to the current specifications for compressive strength testing methods.

Basalt fibers were added to the microwave-absorbing concrete specimens. For the OC, MAC, MCMS, MAB, and MAMB standard specimens for compressive strength and flexural strength (Figure 3d), three parallel specimens were prepared for each type, and the experimental results were averaged. The specimens underwent 50 cycles of freeze–thaw testing, and the strength variations before and after the freeze–thaw cycles were recorded for different structures. The experimental procedure followed the guidelines specified in JTG 3420-2020 [26], with a loading rate of 0.05 to 0.08 MPa/s for flexural strength testing.

2.3.4. Durability of Different Structures

In this study, a concrete abrasion testerYA-2000 (Tianjin Dongzheng Measurement and Control Technology Development Co., Ltd., Tianjing, China) was used. The specimens were prepared with dimensions of 15 cm × 15 cm × 15 cm. After curing under standard conditions for 28 days, the specimens were dried in an oven for 12 h until they reached a constant weight. Three parallel specimens were fabricated for each type, and the experimental results were averaged. Each structure was divided into two groups: one with 0.2% basalt fiber content and the other without basalt fibers (ordinary specimens). Three specimens were prepared for each group, and the results were averaged. After wiping off the surface dust from the specimens, they were placed on the turntable of the abrasion tester. The specimens were subjected to 30 rotations under a load of 200 N using a blade. The initial mass (m_1) was measured after wiping off the surface dust. Subsequently, the specimens were subjected to an additional 60 rotations under a load of 200 N, and the remaining mass (m_2) was measured after wiping off the surface dust. After completing the tests, the wear volume of each pavement structure specimen was calculated using Equation (1). The abrasion resistance of the specimens was characterized using the specific wear volume (Gc) per unit area.

$$G_c = \frac{m_1 - m_2}{A} \quad (1)$$

In Equation (1), Gc represents the specific wear amount per unit area (kg/m^2); m1 denotes the initial mass of the specimen (kg); m_2 refers to the mass of the specimen after wear (kg); and A represents the worn area of the specimen (m_2), where A = 0.0125 m^2 in this experiment.

The standard compressive strength specimens of different structures were placed in a freezer and frozen at a temperature of −20 °C. After 3 h, the specimens were removed from the freezer and heated in a microwave oven for 5 min. Then, the specimens were returned to the freezer for freezing. This cycle was repeated 40 times. Every 10 cycles, the water absorption rate of one specimen was tested and determined using Equation (2). Before measuring the water absorption rate of the specimens, they needed to be dried at 60 °C for 8 h and then at 80 °C for 8 h. The dried specimens were fully immersed in water, and the change in the water absorption rate over time was measured. The water absorption test was stopped after 6 h, and the specimens were air-dried at room temperature before continuing the cycle.

$$\alpha = \frac{m_t - m_0}{\rho_w V_0} \times 100\% \quad (2)$$

In Equation (2), α represents the unit volume water absorption rate. The variables are defined as follows: m_t: the mass of the specimen after a certain immersion time, measured in grams (g); m_0: the mass of the specimen after drying, measured in grams (g); ρ_w: the density of water, measured in grams per cubic centimeter (g/cm^3); V_0: the volume of the specimen before immersion, measured in cubic centimeters (cm^3).

For this experiment, freeze–thaw cycle testing was conducted on the pavement specimens with and without fiber reinforcement using the rapid freezing method. The relative dynamic modulus of elasticity of the specimens was calculated every 25 cycles until reaching 200 cycles. Each group consisted of three specimens, and the average of the test results was taken.

2.3.5. SEM

A Zeiss GeminiSEM 360 scanning electron microscope (SEM, Carl Zeiss (Shanghai) Management Co., Ltd, Shanghai, China) was utilized. The coating materials used for the SEM sample preparation were gold (Au) and platinum (Pt). The acceleration voltage ranged from 0.02 kV to 30 kV, and the probe current ranged from 3 pA to 20 nA. MAB specimens were prepared, and after curing under standard conditions for 28 days, they were analyzed via SEM. The scanning electron microscope was used to study the cement paste and the fiber–cement paste interface transition zone.

3. Results and Discussion

3.1. Heating Effect

The experimental results, as shown in Figure 4a,b, demonstrate the microwave heating of the specimens. The analysis revealed that the heating rate of the top surface of different specimen structures gradually increased with the heating time. The ascending order of heating effectiveness among the four pavement structures is MCMM > MCMS > MAC > OC. The presence of shielding layers in two pavement structures enhances their heating performance compared to the ordinary absorbing concrete pavement structure. Specifically, MCMS and MCMM exhibited approximately 17.2% and 27.1% higher heating rates, respectively, compared to MAC. Although MAC demonstrated favorable heating performance, its microwave thermal conversion efficiency was slightly lower than that of the other two pavement structures with microwave shielding layers. The application of a conductive metal mesh shielding layer and magnetite sand shielding layer significantly improves the microwave thermal conversion efficiency [27]. This indicates that both shielding layers effectively shield and reflect microwaves, reducing microwave leakage to the underlying pavement structure and enhancing microwave heating efficiency [28].

3.2. De-Icing Effect

Due to the lower compressive strength of the conductive metal mesh structure, magnetite sand was chosen as the shielding layer. In this experiment, the MCMS structure was selected to test the de-icing effect. The experimental results were compared with those for the MAC structure.

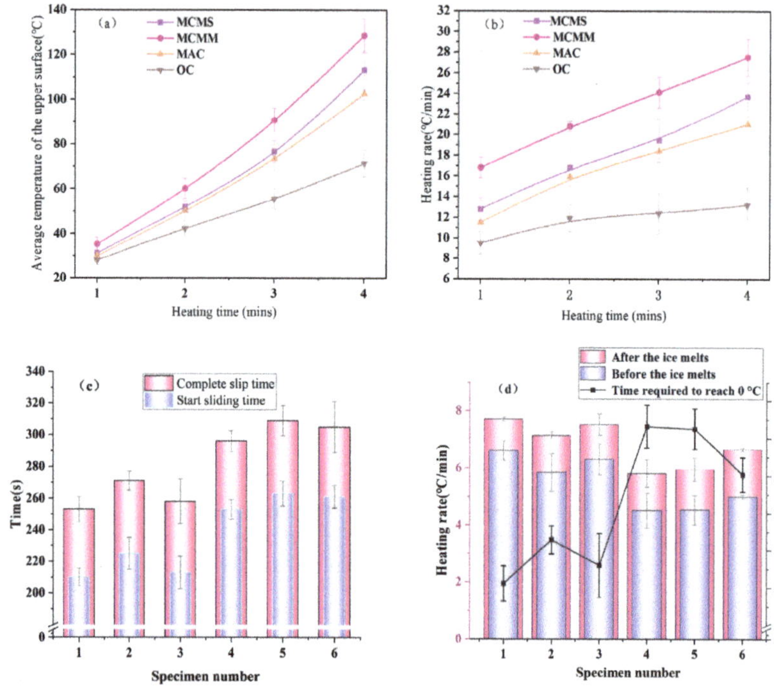

Figure 4. Experimental results for different structural performance. (**a**) Average surface temperature of the upper layer of the specimen; (**b**) heating rate curve of specimen; (**c**) indoor de-icing effect; (**d**) outdoor de-icing effect.

3.2.1. Indoor De-Icing Test

The indoor experimental results are shown in Figure 4c. Specimens 1–3 represent the MCMS structure, while specimens 4–6 represent the MAC structure.

From Figure 4c, it can be observed that the relative sliding time of the ice layer for MCMS and MAC specimens occurred at around 220 s and 260 s, respectively. The complete detachment of the ice layer from the specimen surface took place at approximately 260 s for MCMS and 300 s for MAC. Based on the indoor de-icing test results, the de-icing efficiency of the MCMS structure is approximately 14% higher compared to that of the MAC structure. Additionally, the MCMS structure is capable of reflecting microwaves that are not absorbed by the absorbing concrete. The magnetite sand reabsorbs the reflected microwaves and converts them into heat, thereby improving the microwave thermal conversion efficiency.

3.2.2. Outdoor De-Icing Test

The outdoor de-icing test results are shown in Figure 4d. The analysis revealed that under an initial temperature of −20 °C, it takes approximately 193 s for the surface temperature of the MCMS structure specimen to rise to 0 °C, with a heating rate of 6.25 °C/min. However, it takes approximately 257 s for the surface temperature of the MAC structure specimen to reach 0 °C, with a heating rate of 4.68 °C/min. In the outdoor de-icing test, the heating rate of the specimens was significantly lower compared to that in the indoor de-icing test. This phenomenon can be attributed to the heating conditions in the indoor de-icing test, where the specimens were placed inside a microwave cavity. The microwaves that were not absorbed by the specimens underwent reflection within the cavity and were reabsorbed by the specimens. Additionally, the microwave cavity is a closed environment, which reduces heat loss. As a result, the heating rate of the specimens in the indoor de-icing test was much higher than that in the outdoor de-icing test. Furthermore, the presence of ice layers on the specimen surfaces also had a significant impact on the heating rate.

Due to the stronger heating performance of the MCMS structure, the time required for the surface temperature to rise to 0 °C was shorter, and the heating rate was faster. After the ice layer on the surface of the MCMS structure specimen began to melt, the heating rate was approximately 7.45 °C/min, while for the MAC structure specimen, it was around 6.13 °C/min. The heating rates of both pavement structure specimens significantly increased after the ice layer started to melt. This is because the melting ice layer produced liquid water, which can better absorb microwaves and generate heat, leading to an increase in the surface temperature rise rate of the specimens. However, the heating performance of the MCMS structure was still superior to that of the MAC structure.

3.3. Mechanical Properties

3.3.1. Undoped Basalt Fiber

The compressive strength of four types of specimens, namely, OC, MAC, MCMM, and MCMS, was tested, and the specific experimental results are shown in Figure 5a. As shown in Figure 5a, the compressive strength of the four specimens can be ranked from high to low as MCMS ≈ MAC > OC > MCMM. The compressive strength of the MCMS and MAC specimens increased by 8.0% and 9.0%, respectively, compared to that of the OC specimen. However, the compressive strength of the MCMM specimen decreased by 14.1% compared to that of the OC specimen. The significant decrease in the compressive strength of the MCMM specimen compared to that of the MCMS and MAC specimens is due to the presence of the conductive metal mesh at the interlayer interface, which reduces the bond strength between the two layers of concrete, and thus decreases its compressive strength. The compressive strength test results of the MAC specimens have a strong correlation with their structural characteristics. This is because the concrete containing magnetite has been fully vibrated and has a small thickness of the absorbing layer. Moreover, magnetite has higher hardness and strength than ordinary limestone aggregate, which can cause magnetic blocking during the compressive strength test process and thus exhibit higher compressive strength. Magnetite sand was chosen as a shielding layer material for experimental purposes.

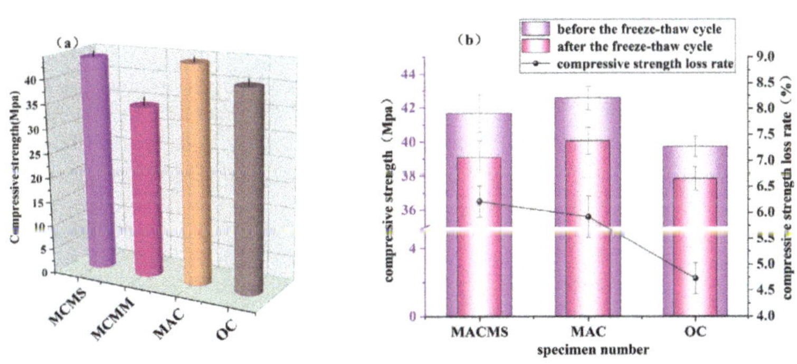

Figure 5. Mechanical strength test of specimens without basalt fiber. (**a**) Compressive strength; (**b**) compressive strength before and after freeze–thaw cycles.

The compressive strength of different absorbing concrete structures was tested for their resistance to multiple "microwave freeze–thaw" cycles, and the experimental results are shown in Figure 5b. As shown in Figure 5b, the compressive strength of MCMS specimens decreased by 2.59 MPa, or 6.22%, after 50 freeze–thaw cycles; and the compressive strength of MAC specimens decreased by 2.52 MPa, or 5.92%, after 50 freeze–thaw cycles; and the compressive strength of OC specimens decreased by 1.88 MPa, or 4.73%, after 50 freeze–thaw cycles. From the perspective of the compressive strength loss rate,

MCMS > MAC > OC. The volume expansion and contraction degree between different aggregates in the road structure specimens containing magnetite was different, and the heating rate of magnetite was greater than that of cement mortar and limestone aggregate. In addition, the heating performance of MCMS specimens was slightly stronger than that of OC specimens, and the shielding layer contained magnetite sand and hematite aggregate, which caused large differences in internal aggregate volume changes, and thus resulted in the fastest decrease in compressive strength for this specimen. Chen et al. [29] studied the mechanical strength of polyurethane elastomers under different environmental conditions and curing times for performance testing.

3.3.2. Incorporated Basalt Fiber

The results of compressive strength tests before and after 60 cycles of freeze–thaw exposure for concrete specimens of different structural types are shown in Figure 6a. The results of flexural strength tests are presented in Figure 6b.

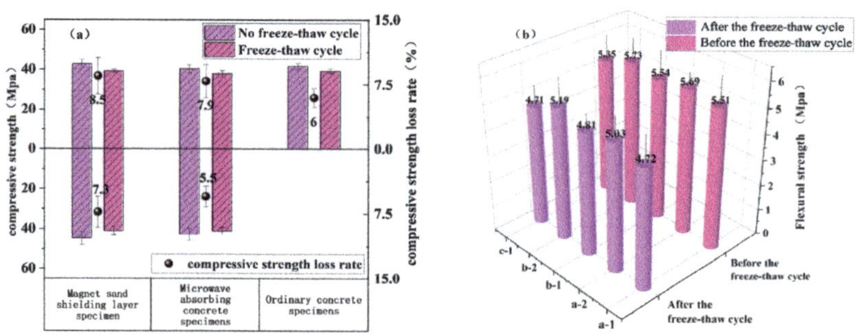

Figure 6. Mechanical properties of specimens before and after freeze–thaw cycles. (**a**) Compressive strength; (**b**) flexural strength.

The experimental results of the microwave-absorbing concrete specimens without basalt fiber are shown above the 0-axis in Figure 6a, while those for the specimens with basalt fiber are shown below the 0-axis. The right axis represents the strength loss rate of the specimen structure type before and after the freeze–thaw cycles. First, we analyze the change in compressive strength of the specimens before and after the freeze–thaw cycles. The compressive strength of the OC specimen after freeze–thawing decreased by 2.51 MPa, or about 6.0%, compared to before freeze–thawing. The compressive strength of the MCMS structure cube specimen decreased by 3.95 MPa, or about 8.5%, compared to before freeze–thawing. The compressive strength of the MAMB specimen decreased by 3.56 MPa, or about 7.3%, after experiencing microwave freeze–thaw cycles. The compressive strength of the MAC structure cube specimen without fiber decreased by 3.55 MPa, or about 7.9%, after freeze–thawing compared to before freeze–thawing. The compressive strength of the MAB specimen decreased by 2.59 MPa, or about 5.5%, after experiencing microwave freeze–thaw cycles. Afroz and Shu et al. [30,31] discovered that basalt fibers exhibit good compatibility with cement. When added to concrete, these fibers can significantly enhance strength and elastic modulus while also increasing toughness and improving durability.

Figure 6b shows that the flexural strength of the OC standard specimen (c-1) after freeze–thawing decreased by 0.64 MPa, or about 12.0%, compared to before freeze–thawing. The flexural strength of the MCMS (a-1) specimen decreased by 0.79 MPa, or about 14.3%, compared to before freeze–thawing. The compressive strength of the MAMB (a-2) specimen decreased by 0.66 MPa, or about 11.6%, after experiencing microwave freeze–thaw cycles. The flexural strength of the MAC standard specimen (b-1) after freeze–thawing decreased by 0.73 MPa, or about 13.2%, compared to before freeze–thawing. The compressive strength of the MAB (B-2) specimen decreased by 0.54 MPa, or about 9.4%, after experiencing microwave freeze–thaw cycles. Nguyen et al. [32] found that incorporating 1.5% basalt fibers into concrete can enhance its compressive and flexural strength.

The compressive and flexural strength loss rates of specimens containing basalt fibers were significantly lower than those of specimens without fibers, and the strength loss rate of specimens without fibers was significantly higher than that of ordinary concrete specimens. This is because during the freeze–thaw cycle, basalt fibers compensate for the defects generated inside the concrete due to volume changes and delay the expansion of cracks; after the specimen cracks, the high elastic modulus of basalt fibers plays a connecting role, so the strength of the specimen can be improved. In addition, basalt fibers can improve the mechanical properties of absorbing concrete under freeze–thaw cycles. Adding basalt fibers can compensate for the strength loss caused by cracks generated inside absorbing concrete due to temperature changes.

3.4. Durability of Different Structures

The results of the abrasion resistance experiment for the OC, MAMB, MAC, MAB, and MCMS specimens, as well as the relative dynamic modulus of elasticity after different numbers of freeze–thaw cycles, are shown in Figure 7a. A total of 200 freeze–thaw cycles were conducted on the specimens, and the relative dynamic modulus of elasticity was recorded every 50 cycles, as shown in Figure 7a. When the specimens of different types underwent 0 freeze–thaw cycles, the water absorption rate varied over time, as shown in Figure 7b. After 20 freeze–thaw cycles, the water absorption rate varied over time, as shown in Figure 7c. After 40 freeze–thaw cycles, the water absorption rate varied over time, as shown in Figure 7d. Ogut et al. [33] found that adding basalt fibers can enhance the durability of concrete.

3.4.1. Wear Resistance

According to the analysis of Figure 7a, the wear loss mass of MCMS and MAC specimens was lower than that of OC specimens. Specifically, compared with ordinary concrete specimens, the wear amount per unit area of MCMS specimens decreased by 10.7%, and that of MAC specimens decreased by 13.6%. The wear resistance of the two types of pavement structure specimens was significantly improved. This change is due to the fact that the absorbing surface layer of MCMS and MAC specimens contains magnetite, whose hardness and strength are superior to those of ordinary limestone gravel aggregate. The hardness of magnetite can reach about 6.5, while that of limestone is only about 4. Therefore, the wear resistance performance of the two pavement structure specimens was much higher than that of ordinary concrete specimens. The wear resistance performance of MCMS and MAC specimens before and after adding basalt fibers also changed. The wear amount per unit area of MAMB decreased by 8.8%, and that of MAB decreased by 9.4%. Therefore, adding basalt fibers to the absorbing concrete pavement structure can effectively improve its wear resistance performance. Alkali-activated materials (AAMs), an alternative to Portland cement, have demonstrated favorable wear performance compared to Portland cement [34]. Ahmed et al. [35] discovered that an optimal content of 1% by weight of cement demonstrated improved abrasion resistance and durability, and met the main requirements of concrete highway pavement, including strength and workability.

3.4.2. Freeze–Thaw Cycle

The analysis of Figure 7a showed that the relative dynamic modulus of elasticity of the four specimens decreased, and the decrease rate of the dynamic modulus of elasticity of the specimens in the early stage of freeze–thaw cycle was relatively slow. When the freeze–thaw cycle had occurred 150 times, the decrease rate increased significantly. The relative dynamic modulus of elasticity of MAC specimens decreased at a lower rate than that of MCMS, indicating that MAC has better frost resistance than MCMS. The frost resistance of MCMS and MAC specimens was improved after adding basalt fibers. The reason for this is that adding basalt fibers to concrete can effectively fill voids, and microwave-absorbing concrete has a denser internal structure. Therefore, the relative dynamic modulus of elasticity of specimens with fibers added decreased slowly in the early stage of freezing and thawing. As hydration inside the concrete progresses, damage caused by freezing and thawing gradually accumulates, and the dynamic modulus of elasticity of specimens decreases rapidly in the later stage. Li et al. [36] analyzed the properties and microstructure of concrete after adding basalt fibers to it. They found that basalt fibers can improve the crack resistance and freeze–thaw resistance of concrete.

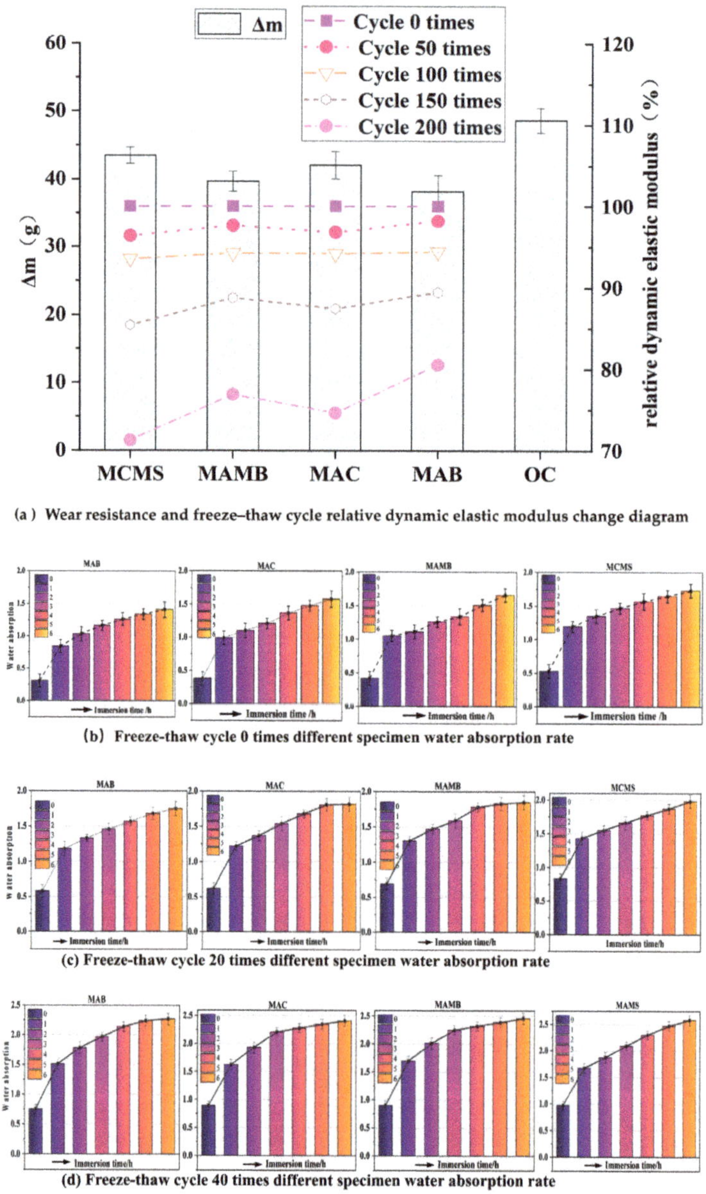

Figure 7. Durability of basalt-fiber-reinforced cement specimens.

3.4.3. Water Absorption

According to the analysis of Figure 7b–d, the water absorption rate of the four structural specimens changed consistently over time, with a significant increase in water absorption rate within the first hour of immersion, followed by a slower increase in water absorption rate over the next five hours. Under the same freeze–thaw cycle conditions, MCMS had a higher water absorption rate than MAC. This may be due to the fact that the lower layer of MCMS specimens contains magnetite and hematite, which have different volume changes from cement materials under microwave freeze–thaw cycle conditions, resulting in cracks and pores and thus increasing the water absorption rate. In addition, the water absorption rate of MAMB specimens decreased after adding basalt fibers. The 6-h water absorption rates under 0, 20, and 40 microwave freeze–thaw cycles decreased by 13.9%, 13.1%, and 12.0%, respectively. Similarly, the 6-h water absorption rate of MAB specimens decreased by 10.8%, 9.9%, and 8.3% under 0, 20, and 40 microwave freeze–thaw cycles, respectively. Therefore, adding basalt fibers to MAC can effectively reduce internal pores in concrete and lower the water absorption rate of pavement structures, thereby reducing cracks caused by liquid water freezing in low-temperature environments. Cui et al. [37] investigated the effects of material composition, porosity, and solute factors on the water absorption of soft rock in their study. Zhang et al. [38] examined the influence of silica fume admixture on the water absorption of hydrophobic concrete.

3.5. SEM

SEM experiments were conducted on specimens containing basalt fibers, and the results are shown in Figure 8.

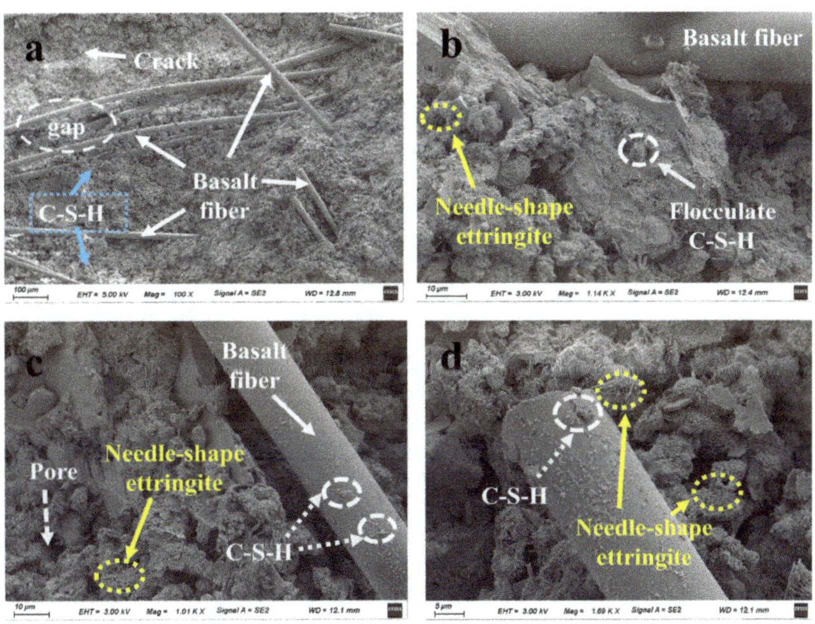

Figure 8. SEM images of basalt-fiber-doped cement specimens after 28 days of curing. (**a**) The microstructure of basalt fiber cement matrix samples at 28 d observed via scanning electron microscopy at a scale of 100 μm; (**b**) the microstructure of basalt fiber cement matrix samples at 28 d observed via scanning electron microscopy at a scale of 10 μm; (**c**) the microstructure of basalt fiber cement matrix samples at 28 d observed via scanning electron microscopy at a scale of 10 μm; (**d**) the microstructure of basalt fiber cement matrix samples at 28 d observed via scanning electron microscopy at a scale of 5 μm.

According to the analysis of Figure 8, the basalt fibers were not tightly connected to the cement slurry, and no chemical reaction occurred between the basalt fibers and the cement slurry. Cement hydration produces flocculent and needle-shaped hydration products. There are gaps between the hydration products of cement hydration, and basalt fibers can fill these gaps, increase the compactness of cement specimens, and enhance the impermeability and durability of cement. C-S-H hydration products of cement hydration adhere to the surface of basalt fibers, increasing the roughness of the surface of basalt fibers. At the same time, a hard shell can be formed on the surface of basalt fibers, which can better transmit stress and improve the overall mechanical properties of cement concrete [22,39].

4. Conclusions

1. After incorporating a microwave shielding layer into microwave-absorbing cementitious concrete structures, the heating and temperature rise effects of MCMM and MCMS structures were improved by 27.1% and 17.2%, respectively, compared to those of the MAC structure. However, the compressive strength of MCMM and MCMS decreased by 14.1% and increased by 8%, respectively, compared to that of OC.
2. MCMS exhibited a 14% improvement in indoor de-icing effectiveness compared to the MAC structure. In outdoor de-icing experiments, by comparing the rate of temperature rise before and after ice melting and the time required for the specimen surface to reach 0 °C, MCMS consistently demonstrated superior performance. Therefore, considering the comprehensive comparison, magnetite sand material was selected as the microwave shielding layer material for microwave de-icing of cementitious concrete pavement to improve de-icing efficiency.
3. After incorporating basalt fibers into cementitious concrete specimens and subjecting them to freeze–thaw cycles, the compressive strength loss rate of MCMS decreased by 1.2%, and the flexural strength attenuation decreased by 2.7%. The strength loss rate of MAB decreased by 2.4%, and the flexural strength rate decreased by 3.8%. Moreover, the addition of basalt fibers also led to certain improvements in the durability performance of cementitious concrete specimens.
4. The microstructure between basalt fibers and the cement matrix was observed and analyzed using SEM, providing microscopic evidence for the enhancement of mechanical strength and durability of cementitious concrete by incorporating basalt fibers. Therefore, a volume fraction of 0.2% of basalt fibers was selected to enhance the mechanical strength and durability of microwave-absorbing cementitious concrete pavement.
5. In this study, MCMB and MAB specimens were formed at room temperature. Further research is needed to investigate the performance of specimens formed with the addition of basalt fibers under low-temperature conditions. Additionally, a comprehensive evaluation method and system for the performance of MAC pavement have not yet been established. Future studies could focus on scientifically and systematically researching microwave de-icing technology considering climate conditions and economic factors.

Author Contributions: Conceptualization, A.J. and X.W.; methodology, Z.S.; validation, A.J., J.Z. and J.R.; formal analysis, Z.S.; investigation, J.Z. and J.R.; resources, A.J. and X.W.; data curation, Z.S. and J.Z.; writing—original draft preparation, X.W. and J.R.; writing—review and editing, A.J. and Z.S.; supervision, A.J.; project administration, A.J. All authors have read and agreed to the published version of the manuscript.

Funding: This research received no external funding.

Institutional Review Board Statement: Not applicable.

Informed Consent Statement: Not applicable.

Data Availability Statement: All data, models, and code generated or used during the study appear in the published article.

Conflicts of Interest: The authors declare no conflict of interest.

References

1. Bardal, K.G.; Jørgensen, F. Valuing the risk and social costs of road traffic accidents—Seasonal variation and the significance of delay costs. *Transp. Policy* **2017**, *57*, 10–19. [CrossRef]
2. Andleeb, Z.; Khawaja, H.; Moatamedi, M. Multiphysics based Analysis of Materials for Roads in Cold Regions to Prevent Ice Adhesion and Low-Temperature Crack Developments. *Int. J. Multiphys.* **2022**, *16*, 437–448.
3. Peng, C.; Hu, Y.D.; You, Z.P.; Yang, H.; Nie, Y.T.; Wu, T.H.; Yang, H.N.; Ou, R.H. Preparation and anti-icing performance of acrylic superhydrophobic asphalt pavement coating with microwave heating function. *Constr. Build. Mater.* **2022**, *344*, 128289. [CrossRef]
4. Theofilatos, A. Incorporating real-time traffic and weather data to explore road accident likelihood and severity in urban arterials. *J. Saf. Res.* **2017**, *61*, 9–21. [CrossRef]
5. Deng, Y.; Wang, X.; Chen, L.; Liu, M.; Gao, M.; Zhao, J. A Study on the Heating and Deicing Performance of Microwave-Absorbing Asphalt Mixtures. *Materials* **2023**, *16*, 1051. [CrossRef]
6. Zhou, J.; Lai, S.; Xu, X.; Chen, Y.; Chu, W.; Gao, Y. R&D of Equipment for Deicing by Thermal Water-jet and Mechanical Deicing Method. In Proceedings of the 2nd International Conference on Applied Mechanics, Materials and Manufacturing (ICAMMM 2012), Changsha, China, 17–18 November 2012; pp. 1288–1293.
7. Gao, J.; Sha, A.; Wang, Z.; Tong, Z.; Liu, Z. Utilization of steel slag as aggregate in asphalt mixtures for microwave deicing. *J. Clean. Prod.* **2017**, *152*, 429–442. [CrossRef]
8. Zhao, J.; Wang, X.; Xin, L.; Ren, J.; Cao, Y.; Tian, Y. Concrete pavement with microwave heating enhancement functional layer for efficient de-icing: Design and case study. *Cold Reg. Sci. Technol.* **2023**, *210*, 103846. [CrossRef]
9. Micheli, D.; Pastore, R.; Vricella, A.; Morles, R.B.; Marchetti, M.; Delfini, A.; Moglie, F.; Primiani, V.M. Electromagnetic characterization and shielding effectiveness of concrete composite reinforced with carbon nanotubes in the mobile phones frequency band. *Mater. Sci. Eng. B-Adv. Funct. Solid-State Mater.* **2014**, *188*, 119–129. [CrossRef]
10. Micheli, D.; Vricella, A.; Pastore, R.; Delfini, A.; Morles, R.B.; Marchetti, M.; Santoni, F.; Bastianelli, L.; Moglie, F.; Primiani, V.M.; et al. Electromagnetic properties of carbon nanotube reinforced concrete composites for frequency selective shielding structures. *Constr. Build. Mater.* **2017**, *131*, 267–277. [CrossRef]
11. Wang, Z.; He, Z.; Wang, Z.; Ning, M. Utilization of magnetite as microwave absorber to prepare microwave-heatable aggregate for deicing in cementitious composite. *Constr. Build. Mater.* **2019**, *227*, 116664. [CrossRef]
12. Wang, Z.; He, Z.; Wang, Z.; Ning, M. Microwave Deicing of Functional Pavement Using Sintered Magnetically Separated Fly Ash as Microwave-Heating Aggregate. *J. Mater. Civ. Eng.* **2019**, *31*, 04019127. [CrossRef]
13. Liu, H.-I.; Shi, J.-J.; Qu, H.-Q.; Ding, D.-X. Feasibility of using recycled CRT funnel glass as partial replacement of high density magnetite sand in radiation shielding concrete. *Trans. Nonferrous Met. Soc. China* **2019**, *29*, 831–839. [CrossRef]
14. Guo, H.; Wang, Z.; Huo, J.; Wang, X.; Liu, Z.; Li, G. Microwave heating improvement of asphalt mixtures through optimizing layer thickness of magnetite and limestone aggregates. *J. Clean. Prod.* **2020**, *273*, 123090. [CrossRef]
15. Trigos, L.; Gallego, J.; Ignacio Escavy, J. Heating potential of aggregates in asphalt mixtures exposed to microwaves radiation. *Constr. Build. Mater.* **2020**, *230*, 117035. [CrossRef]
16. Lu, S.; Bai, E.; Xu, J.; Chen, J. Research on electromagnetic properties and microwave deicing performance of carbon fiber modified concrete. *Constr. Build. Mater.* **2021**, *286*, 122868. [CrossRef]
17. Liu, M.Y.; Wang, X.C.; Deng, Y.Y.; Guo, Y.C.; Zhao, J.; Li, M.X. Study on Microwave Deicing Efficiency of Microwave-Absorbing Concrete Pavements and Its Influencing Factors. *Materials* **2022**, *15*, 8923. [CrossRef]
18. Guan, B.; Liu, J.; Zhao, H.; Wu, J.; Liu, J.; Yang, F. Investigation of the Microwave Absorption of Asphalt Mixtures Containing Magnetite Powder. *Coatings* **2019**, *9*, 813. [CrossRef]
19. Wang, Z.; Bai, E.; Huang, H.; Wang, T.; Sun, H. Study on the electromagnetic property and microwave heating efficiency of concrete with magnetite aggregate. *Constr. Build. Mater.* **2022**, *342*, 128080. [CrossRef]
20. Zhao, Y.; Liu, X.M.; Wei, D.M. Asphalt mixture with nucleus-shell structural microwave enhanced aggregate: Structure design, preparation and performance. *Case Stud. Constr. Mater.* **2023**, *19*, e02248. [CrossRef]
21. Chung, S.-i.; Kim, P.K.; Ha, T.-g. High-performance transparent electromagnetic interference shielding film based on metal meshes. *J. Micromech. Microeng.* **2023**, *33*, 035002. [CrossRef]
22. Li, Z.; Chen, F.; Ren, J.; Chen, Z. Experimental Study on the Impermeability and Micromechanisms of Basalt Fiber-Reinforced Soil-Cement in Marine Environments. *Coatings* **2023**, *13*, 532. [CrossRef]
23. Wang, Z.H.; Bai, E.; Ren, B.; Liu, C.J.; Liu, J.L.; Huang, H. Microwave heating efficiency and frost resistance of concrete modified with powder absorbing materials. *Constr. Build. Mater.* **2023**, *379*, 131145. [CrossRef]
24. Fu, C.; Liu, K.; Liu, Q.; Xu, P.; Dai, D.; Tong, J. Exploring directional energy conversion behavior of electromagnetic-based multifunctional asphalt pavement. *Energy* **2023**, *268*, 126573. [CrossRef]
25. *JTG D40-2011*; Ministry of Communications of the People's Republic of China Specifications for Design of Highway Cement Concrete Pavement. Ministry of Communications: Beijing, China, 2011.
26. *JTG 3420-2020*; Ministry of Communications of the People's Republic of China Testing Methods of Cement and Concrete for Highway Engineering. Ministry of Communications: Beijing, China, 2020.

27. Merizgui, T.; Hadjadj, A.; Kious, M.; Gaoui, B. Impact of Temperature Variation on the Electromagnetic Shielding Behavior of Multilayer Shield for EMC Applications. *Rev. Compos. Mater. Av.-J. Compos. Adv. Mater.* **2019**, *29*, 363–367. [CrossRef]
28. Qiu, H.; Chen, H.; Wu, Y.; Wu, J.; Liu, Y.; Lin, Y. Heating Characteristics and Deicing Properties of Magnetite Mortar Microwave-Absorbing Layer on Concrete Pavement. *J. Mater. Civ. Eng.* **2023**, *35*, 15729. [CrossRef]
29. Chen, Q.; Wang, C.H.; Li, Y.W.; Feng, L.; Huang, S. Performance Development of Polyurethane Elastomer Composites in Different Construction and Curing Environments. *Constr. Build. Mater.* **2023**, *365*, 130047. [CrossRef]
30. Afroz, M.; Patnaikuni, I.; Venkatesan, S. Chemical durability and performance of modified basalt fiber in concrete medium. *Constr. Build. Mater.* **2017**, *154*, 191–203. [CrossRef]
31. Shu, Y.; Zhang, J. Effect of Basalt Fiber Content and Length on the Strength and Crack Development of Polyvinyl Alcohol/Basalt Hybrid Fiber-Reinforced Cement Soil. *Polymers* **2023**, *15*, 2146. [CrossRef]
32. Ngoc, L.N.; Van, H.L. Mechanical and Shrinkage Behavior of Basalt Fiber Reinforced Ultra-High-Performance Concrete. *Int. J. Geomate* **2021**, *20*, 28–35.
33. Ogut, R.; Demir, A. The Effect of the Basalt Fiber on Mechanical Properties and Corrosion Durability in Concrete. *Arab. J. Sci. Eng.* **2023**, *48*, 5097–5114. [CrossRef]
34. Shagnay, S.; Velasco, F.; del Campo, A.; Torres-Carrasco, M. Wear behavior in pastes of alkali-activated materials: Influence of precursor and alkali solution. *Tribol. Int.* **2020**, *147*, 106293. [CrossRef]
35. Ahmed, N.Y.; Alkhafaji, F.F. Enhancements and mechanisms of nano alumina (Al_2O_3) on wear resistance and microstructure characteristics of concrete pavement. *IOP Conf. Ser. Mater. Sci. Eng.* **2020**, *871*, 012001. [CrossRef]
36. Li, Z.; Shen, A.; Zeng, G.; Chen, Z.; Guo, Y. Research progress on properties of basalt fiber-reinforced cement concrete. *Mater. Today Commun.* **2022**, *33*, 104824. [CrossRef]
37. Cui, G.; Zhou, C.; Liu, Z. Prediction Method of Water Absorption of Soft Rock Considering the Influence of Composition, Porosity, and Solute Quantitatively. *Appl. Sci.* **2022**, *12*, 104824. [CrossRef]
38. Zhang, C.; Zhang, S.; Yu, J.; Kong, X. Water absorption behavior of hydrophobized concrete using silane emulsion as admixture. *Cem. Concr. Res.* **2022**, *154*, 106738. [CrossRef]
39. Xu, X.; Cui, S.; Xu, L.; Zeng, G.; Wang, X. Effect of different fiber sizing on basalt fiber-reinforced cement-based materials at low temperature: From macro mechanical properties to microscopic mechanism. *Constr. Build. Mater.* **2023**, *392*, 131773. [CrossRef]

sustainability

MDPI

Article

Comprehensive Laboratory Evaluation of Crack Resistance for an Asphalt Rubber Stress-Absorbing Membrane Interlayer (AR-SAMI)

Ping Li [1], Wenju Peng [1,2,*], Shuaituan Tian [3], Zhaohui Liu [1], Junbin Liu [4] and Shende Liu [1]

1 School of Traffic & Transportation Engineering, Changsha University of Science & Technology, Changsha 410114, China; li_ping@csust.edu.cn (P.L.)
2 School of Civil Engineering, Hubei Engineering University, Xiaogan 432000, China
3 Beijing Super-Creative Technology Co., Ltd., Beijing 100094, China
4 Gezhouba Group Transportation Investment Co., Ltd., Wuhan 430030, China
* Correspondence: pwj1112@foxmail.com

Abstract: Reflective cracking is a common distress of semi-rigid base asphalt pavements and overlay pavement projects. An asphalt rubber stress-absorbing membrane interlayer (AR-SAMI) prepared by waste tires is an effective engineering solution for treating reflective cracking. This method can also reduce black pollution. However, there is no unified test method and index for crack resistance evaluation. In this work, AR-SAMIs with different air voids and gradations were investigated. A small beam bending test (BBT) at −10 °C and 15 °C, crack expansion SCB (CE-SCB) test, low-temperature SCB (LT-SCB) test, and Overlay Test (OT) were performed to evaluate the crack resistance of AR-SAMI comprehensively. Statistical analysis was also performed. Results showed that the crack resistance of AR-SAMI improved as the air voids decreased. The crack resistance of 10-A gradation with more fine aggregate was excellent. However, the AR-SAMI with more coarse aggregate had better crack extension resistance under the condition of pre-existing cracks. There are differences in the evaluation results of different test methods due to the various evaluation focus. The −10 °C BBT, CE-SCB, and OT were recommended to evaluate the crack resistance comprehensively. Research results can guide the evaluation method or index selection of crack resistance and the optimization of AR-SAMI mixture design under different working conditions.

Keywords: crack resistance; asphalt rubber stress-absorbing membrane interlayer (AR-SAMI); reflective cracking; air voids; gradation

Citation: Li, P.; Peng, W.; Tian, S.; Liu, Z.; Liu, J.; Liu, S. Comprehensive Laboratory Evaluation of Crack Resistance for an Asphalt Rubber Stress-Absorbing Membrane Interlayer (AR-SAMI). *Sustainability* **2023**, *15*, 8982. https://doi.org/10.3390/su15118982

Academic Editor: Antonio D'Andrea

Received: 1 May 2023
Revised: 23 May 2023
Accepted: 29 May 2023
Published: 2 June 2023

1. Introduction

Reflective cracking is a common type of distress in asphalt pavement caused by existing cracks in the underlying layer [1,2]. Semi-rigid base asphalt pavements, composite road (or bridge) pavements, and pavement maintenance of overlay projects have potential risks of reflective cracking. Reflective cracking can cause fissuration, loosening, and other distress under the coupling effect of multiple factors, such as water, loading, and temperature. It can significantly reduce pavement performance and directly affect the service life of the pavement [3,4]. Therefore, how to effectively mitigate the occurrence of reflective cracking and improve the quality of pavement engineering has become one of the urgent technical problems for pavement construction and maintenance [5,6].

The commonly used techniques for treating reflective cracking include the following three types: laying geosynthetic materials, adding an unbound aggregate layer between the surface layer and the base layer, and adding a stress-absorbing membrane interlayer (SAMI) [7–11]. Among them, the SAMI has been widely used as an effective method to slow the expansion of reflective cracking [12–14]. Due to its special functional requirements, a SAMI generally uses polymer modified asphalt, crumb rubber modified asphalt (CRMA), and high-viscosity asphalt [15]. CRMA had excellent properties (including high and low temperature performance and insensitivity to temperature changes) and economic performance. More importantly, the crumb rubber prepared from waste tires has excellent environmentally friendly properties [16,17]. It meets the requirements of sustainable social development. Thus, the asphalt rubber stress-absorbing membrane interlayer (AR-SAMI) has been widely used [3].

Air void and gradation are key elements of mixture design and significantly impact the performance of CR mixture or AR-SAMI. For example, Asgharzadeh et al. [18] showed that the fatigue life could be improved by 9.4 to 18.2 times by reducing the air void from 4.0% to 2.0%. Aliha et al. [19] demonstrated that by increasing the air void from 3% to 7%, the fracture toughness of AR decreased significantly. Also, the gradation significantly affected the performance of CR mixture or AR-SAMI [20]. Meanwhile, existing research has further optimized the performance of AR-SAMI and improved its application effect. For example, the construction temperature of the crumb rubber asphalt mixture was high due to its high viscosity. Pan et al. [21] used warm mix technology to reduce the compaction temperature by 30 °C. It still had good low-temperature performance, which could be suitable for pavement cracking disposal in cold areas. Zeng et al. [12] used fibers for AR-SAMI modification, which improved its crack resistance and fatigue life on the semi-rigid base. Wang et al. [22] used amorphous poly alpha-olefin (APAO) to modify AR-SAMI. The Overlay Test (OT) results showed that the adhesion performance and crack resistance of the mixture were improved.

Unlike conventional asphalt mixtures, SAMI should have good crack resistance in addition to the traditional pavement performance requirements such as high temperature, low temperature, and water stability. Crack resistance includes the ability to resist crack formation and crack expansion [3]. Numerical simulation techniques have been used to investigate it. Among them, finite element (FE) and discrete element method (DEM) were most commonly used. For example, Thives et al. [23] used Ansys calculations to show that pavement fatigue life could be improved up to eight times using AR-SAMI compared to a regular asphalt mixture under Brazilian climatic conditions. Zeng et al. [12] used FE to verify that AR-SAMI with fibers could significantly reduce the stress concentration in cracks. Pei et al. [24] used DEM to analyze the destructive properties of asphalt pavement overlay on cement pavement, which guided the application of the paving technology under different working conditions. However, due to the complexity of the effects of cracking, it was not easy to accurately construct an analytical model that matched the actual material composition and application conditions. As a result, the numerical simulation had a significant deficiency in material selection and mixture composition optimization design. Therefore, the experimental evaluation methods still had the most direct guidance value for the mixture design of SAMI.

However, there was no unified standard for crack resistance test evaluation. Therefore, scholars have proposed a variety of test evaluation methods. For example, the small beam bending test (BBT) was utilized in China [25]. The semicircular bending (SCB) tests with pre-cut cracks have been set as the standard test by AASHTO, including the crack extension SCB (CE-SCB) test [26] and low-temperature SCB (LT-SCB) test [27]. There are also Overlay Tests (OT) in Texas [22,28] and wheel fatigue tests [29]. Depending on the generation mechanism, the causes of cracks vary significantly under different working conditions. Four major modes of asphalt pavement cracking exist, including thermal, reflection, bottom-up fatigue, and top-down [30]. Due to the different testing principles, the crack resistance performance reflected by each index varies greatly. For example, the low-temperature flexural-tensile strain in BBT is the control index of the Chinese specification, which focuses on characterizing the resistance to crack formation [31]. The SCB tests with pre-cut cracks focus on characterizing the resistance to crack expansion in the presence of cracks [9,32]. The OT test is similar to the fatigue test and characterizes the crack resistance degradation of the specimen under fixed deformation [9,33].

To sum up, existing studies have investigated the effect of AR-SAMI on traditional road performance but less on crack resistance. Meanwhile, numerical simulation techniques were generally used in existing studies to evaluate the crack resistance of SAMI. There is a lack of systematic standards for different working conditions when using test methods, as the standards are not unified across countries. Meanwhile, the existing studies generally used a single evaluation technique, which could not comprehensively characterize the comprehensive crack resistance performance of the SAMI. As a result, they could not guide the optimal design of the SAMI. Although AR-SAMI has been initially promoted in the treatment of reflective cracking, the evaluation of its comprehensive anti-cracking performance needs to be further strengthened. The evaluation method and index applicability research especially still need to be deepened.

In this work, three gradation types of AR-SAMI mixtures with two air voids (AV, 2.5% and 4.0%) were designed. The anti-cracking performances of AR-SAMI were evaluated through different temperatures (−10 °C, 15 °C) of BBT, CE-SCB test, LT-SCB test, and OT. Meanwhile, the test results were compared by statistical methods to select the appropriate comprehensive evaluation test methods and indices. This work can provide a reference for the crack resistance evaluation of similar SAMI and can guide the mixture design of AR-SAMI under different working conditions.

2. Materials and Methods

The research flowchart of this study can be depicted in Figure 1.

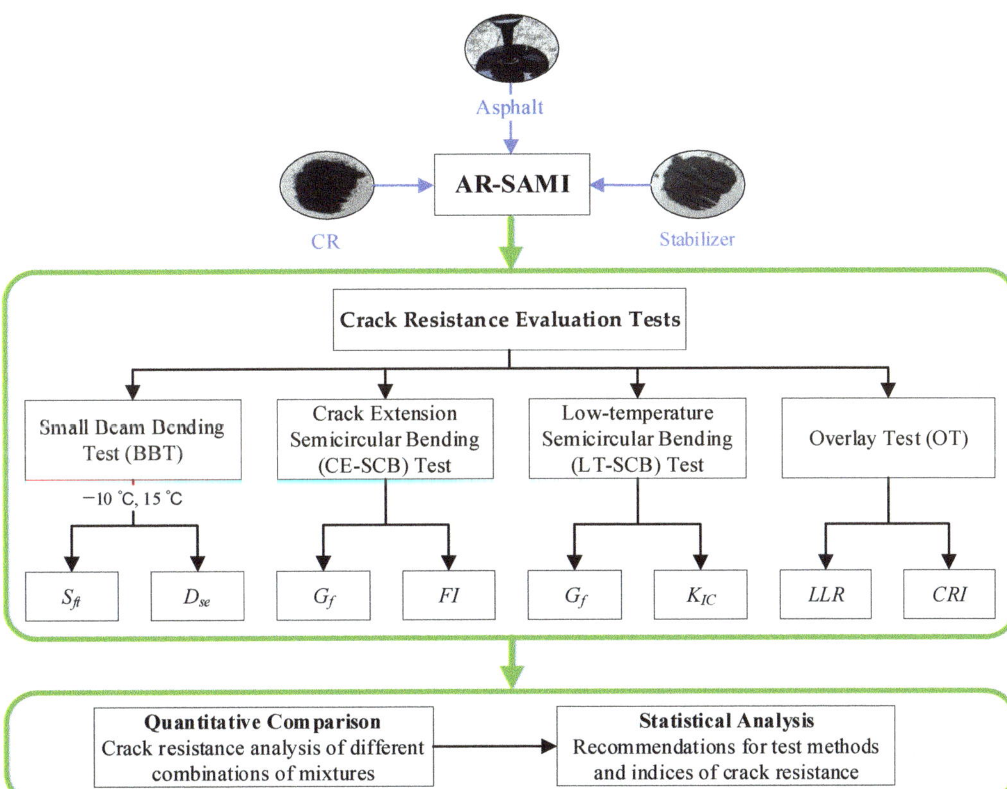

Figure 1. Research flowchart.

2.1. *Crumb Rubber Modified Asphalt*

The crumb rubber modified asphalt was prepared with 70# A base asphalt, crumb rubber, and stabilizer (as shown in Figure 2). The dosage of crumb rubber was 20 wt%. The stabilizer is shown in Figure 2, produced by Guangxi Transportation Science and Technology Group Co., Ltd. (Nanning, China).

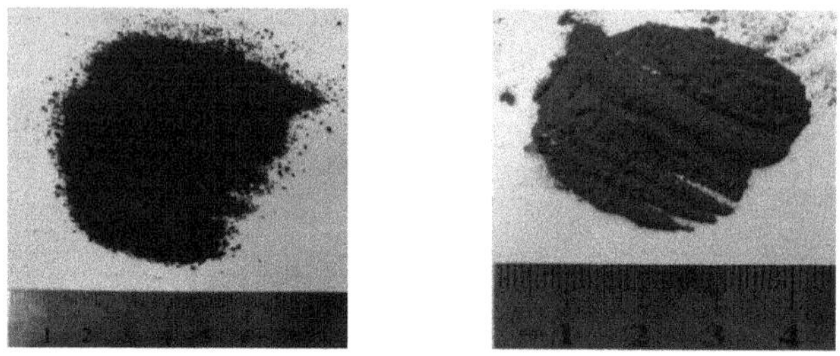

(**a**) (**b**)

Figure 2. Crumb Rubber and stabilizer of crumb rubber modified asphalt. (**a**) Crumb rubber (**b**) Stabilizer.

The technical indices of crumb rubber modified asphalt are shown in Table 1.

Table 1. Technical indices of crumb rubber modified asphalt.

Technical Indices	Units	Test Results	Test Methods [34]
Penetration (25 °C, 5 s, 100 g)	0.1 mm	39.3	T 0604
Ductility (5 °C, 5 cm/min)	cm	11.6	T 0605
Softening point	°C	75.5	T 0606
Brookfield viscosity (180 °C)	Pa·s	2.51	T 0625
Elastic recovery rate (25 °C)	%	94.0	T 0662

2.2. *Aggregate*

The limestone crushed stone was used as coarse and fine aggregate. The main technical indices of aggregate are shown in Table 2. The technical indices meet the performance requirements of highway aggregate [31].

Table 2. Main technical indices of aggregate.

Aggregate Types	Technical Indices		Units	Test Results	Requirements [31]	Test Methods
Coarse aggregate	Crushing value		%	17	≤26	T 0316
	Apparent specific gravity	9.5~13.2 mm	—	2.705	≥2.60	T 0304
		4.75~9.5 mm		2.735		
		2.36~4.75 mm		2.63		
	Bulk specific gravity	9.5~13.2 mm	—	2.676	—	T 0304
		4.75~9.5 mm		2.654		
		2.36~4.75 mm		2.611		
	Water absorption	9.5~13.2 mm	%	0.41	≤2.0	T 0304
		4.75~9.5 mm		1.11		
		2.36~4.75 mm		0.27		
	Flat and elongated particles content	9.5~13.2 mm	%	13.2	≤15	T 0312
		4.75~9.5 mm		10.9	≤15	
		2.36~4.75 mm		7.5	≤20	
	Adhesional degree with aggregate		—	5	≥4	T 0616

Table 2. *Cont.*

Aggregate Types	Technical Indices	Units	Test Results	Requirements [31]	Test Methods
Fine aggregate	Apparent specific gravity	—	2.622	≥2.50	T 0304
	Bulk specific gravity	—	2.620	—	T 0304
	Water absorption	%	0.27	—	T 0304
	Sand equivalent	%	69	≥60	T 0334
	Angularity (Flow time method)	s	32.2	≥30	T 0345

2.3. Mixture Design

There was no national standard for the design of AR-SAMI. Considering AC-10 type was mostly used in the highway project, the AC-10 gradations in JTG F40-2004 [31], DB45/T 1098-2014 [35], and DG/TJ08-2109-2012 [36] were analyzed. Three gradations were determined based on summary results, as shown in Table 3.

Table 3. Test gradations of AR-SAMI.

Gradation Types	Mass Percentage Passing the Following Square-Mesh Sieve (mm)/%								
	13.2	9.5	4.75	2.36	1.18	0.6	0.3	0.15	0.075
10-A	100	91	57	40	27	19	13	9	6
10-B-1	100	95	35	28	21	17	13	10	8
10-B-2	100	98	38	29	23	15	11	8	6

Two air voids of 2.5% and 4.0% were selected. The mixture design was based on the combination of gradation and air voids. The Marshall design method was used to determine the optimal asphalt–aggregate ratio and volume parameters. The test results are shown in Table 4. Among them, voids in the mineral aggregate (VMA) and voids filled with asphalt binder (VFA) comply with the relevant requirements of JTG F40-2004 [31].

Table 4. Mixture design results of each gradation.

Gradation Types	Air Voids/%	Optimal Asphalt–Aggregate Ratio/%	VMA/%	VFA/%	Asphalt Film Thickness/μm
10-A	4.0	5.75	15.78	74.69	9.13
	2.5	6.10	15.11	83.41	9.70
10-B-1	4.0	5.57	15.40	74.07	8.21
	2.5	6.55	15.93	84.38	9.71
10-B-2	4.0	5.89	16.04	75.13	10.46
	2.5	6.46	15.78	84.18	11.51

2.4. Gradation Evaluation by Bailey Method

The Bailey method was used to quantify and analyze the differences between the three gradations.

The following three indicators were used according to the theory of the Bailey method [37]: Coarse Aggregate Ratio (CA), Fine Aggregate Coarse Ratio (FA_C), and Fine Aggregate Fine Ratio (FA_f). The formulae for calculating each index are as follows:

$$CA = \frac{P_{D/2} - P_{PCS}}{100 - P_{D/2}} \tag{1}$$

$$FA_C = \frac{P_{SCS}}{P_{PCS}} \tag{2}$$

$$FA_f = \frac{P_{TCS}}{P_{SCS}} \tag{3}$$

where P is the passing rate of the corresponding sieve, D is Nominal Maximum Aggregate Size (NMAS), Primary Control Sieve (PCS) is 0.22 × D, the Secondary Control Sieve (SCS) is 0.22 × PCS, and the Tertiary Control Sieve (TCS) is 0.22 × SCS.

2.5. *Crack Resistance Evaluation Test Methods*

2.5.1. Small Beam Bending Test (BBT)

The BBTs at −10 °C and 15 °C were conducted with reference to JTG E20-2011 (T 0715-2011) [34]. The specimen size was 250 mm × 30 mm × 35 mm, and the span diameter was 200 mm. The three-point loading tests were carried out by using the universal material testing machine (Shanghai Hualong WDW-100C). The loading rate was 50 mm/min. Three parallel specimens were tested in each group. The mean and standard deviation were calculated separately.

The maximum flexural-tensile strain (S_{ft}) and strain energy density (D_{se}) were calculated. The maximum flexural-tensile strain indicates the maximum strain value of the small beam specimen during the test. The strain energy density means the energy consumed by the specimen during the whole process of loading damage. Among them, the strain energy density is the strain energy per unit volume, which can be calculated by using the envelope area of the σ (stress)-ε (strain) curve during the damage process [38]. The calculation formula is as follows:

$$D_{se} = \int_0^{\varepsilon_B} \sigma d\varepsilon \quad (4)$$

where ε_B is the corresponding strain to the stress peak of the specimen (mm).

2.5.2. Crack Extension Semicircular Bending (CE-SCB) Test

The CE-SCB test was performed at room temperature (25 ± 0.5 °C) using pre-cut crack semicircular specimens in accordance with AASHTO TP 124-16 [26]. The specimen dimensions are shown in Figure 3. The notches were 15 ± 1 mm deep and 2.0 ± 1.0 mm wide. Each group of tests has four parallel specimens. The mean and standard deviation were calculated separately.

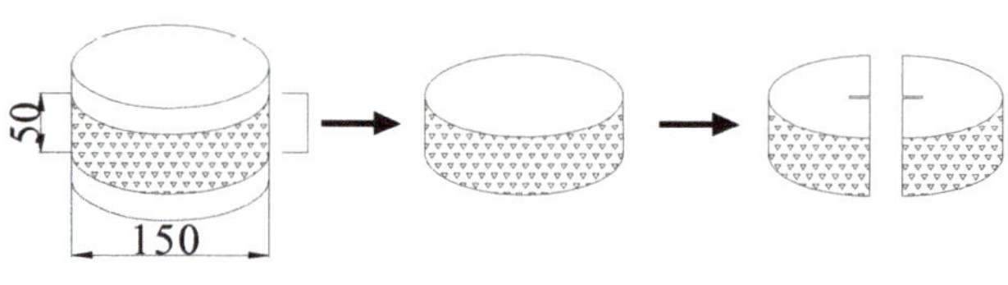

(a)

(b)

Figure 3. Specimen preparation and specimen process of CE-SCB test. (**a**) Schematic diagram of specimen preparation. (**b**) Test loading process.

The fracture energy (G_f) and the flexibility index (FI) could be calculated from the load–displacement curve (as shown in Figure 4). G_f denotes the energy required to create a unit surface area of a crack, and FI is an index intended to characterize the damage resistance of asphalt mixtures. The larger the value of the two indicates the stronger the crack resistance of asphalt mixtures under normal temperature conditions. Among them, FI takes into account the slope of the force–displacement curve m at the later stage of crack expansion (Figure 4). It could indicate the resistance to crack expansion after the formation of cracks in the mixture.

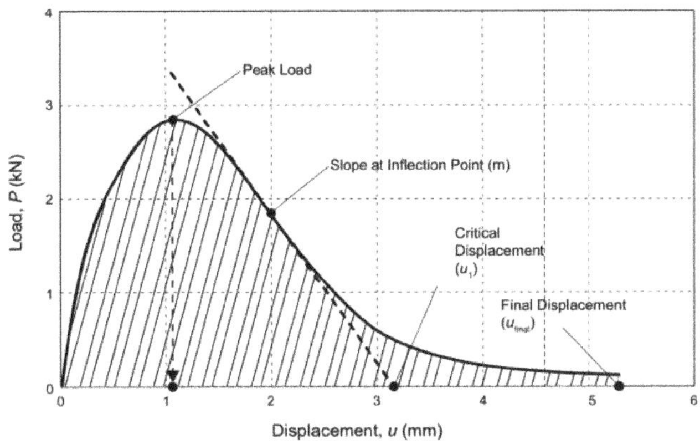

Figure 4. Typical force–displacement curve of the CE-SCB test [26].

2.5.3. Low-Temperature Semicircular Bending (LT-SCB) Test

LT-SCB test used semicircular pre-cut specimens to evaluate the ability of the mixture to resist damage under low-temperature conditions referring to the AASHTO TP 105-13 test method [27]. The specimen preparation of the LT-SCB test was roughly the same as the CE-SCB test. The difference was that the thickness of the cut specimen was adjusted to 24.7 ± 2 mm. The test temperature was −10 °C. Each group of tests had four parallel specimens. The mean and standard deviation were calculated separately.

The fracture energy G_f and fracture toughness K_{IC} (also known as mode I critical stress intensity factor) of the asphalt mixture under low-temperature conditions were calculated with reference to the specification. The higher the index value, the better the resistance of the compound to crack expansion at low temperatures.

2.5.4. Overlay Test (OT)

The OT was performed with reference to the TEX-248-F [39]. The Marshall compaction method (T 0702-2011) [34] was used to form specimens (152.4 mm × 95.3 mm) The cylindrical specimen with a 38.1 ± 0.5 mm thickness was obtained by cutting off the top and bottom parts. Then, both ends of the specimen were cut, and the specimen sample with a width of 76.2 ± 0.5 mm was obtained. The specimen-cutting process and the test procedure are shown in Figure 5.

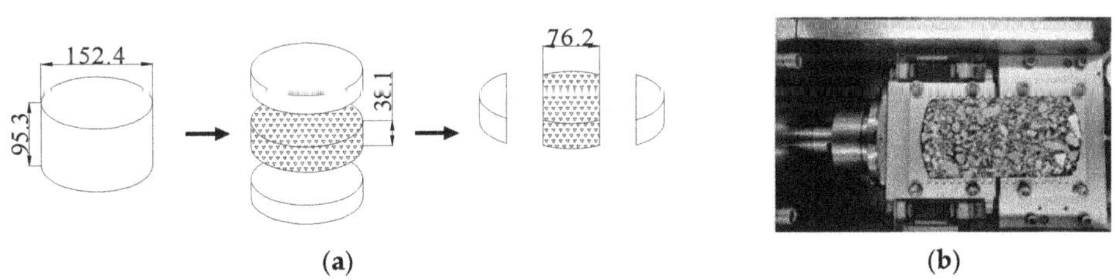

Figure 5. Schematic diagram of OT test. (**a**) Schematic diagram of specimen cutting process. (**b**) OT test process diagram.

The test temperature was 25 ± 0.5 °C. A relaxation period of at least 10 min should be maintained before the test. The test would stop when the maximum load was reduced by 93%. When loading cycles reached 1200, it would also terminate the test. The initial load and final load were recorded. Then, the load loss rate (*LLR*) and crack resistance index (*CRI*) were calculated according to the specification. There were three parallel specimens in each group. The mean and standard deviation were calculated separately.

The OT is a tensile fatigue test. *LLR* indicates the reduction rate of the final load compared with the initial loading. The smaller the value, the better the resistance to damage. *CRI* is an inherent property of the specimen, which is obtained by fitting the power equation of the curve of the load and the number of loading cycles during the loading process. The higher the value, the better the crack resistance of the mixture.

3. Results and Discussion

3.1. Small Beam Bending Test

The BBTs were conducted at −10 °C and 15 °C, respectively. Test results are shown in Figures 6 and 7.

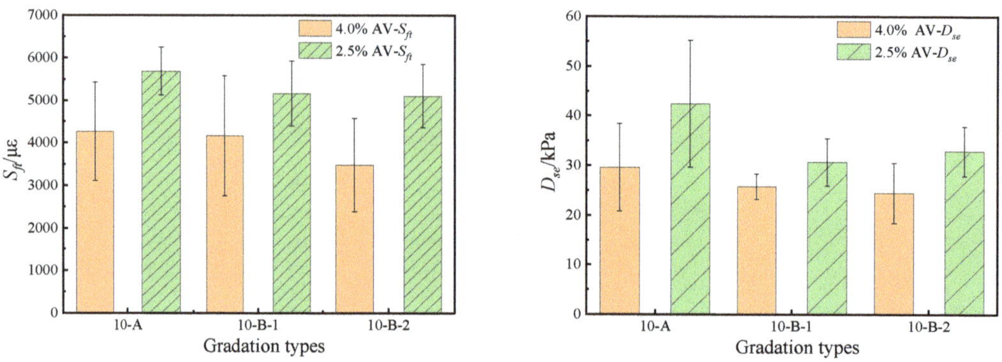

(a) **(b)**

Figure 6. Small beam bending test results of AR-SAMI @ −10 °C. (**a**) S_{ft}. (**b**) D_{se}.

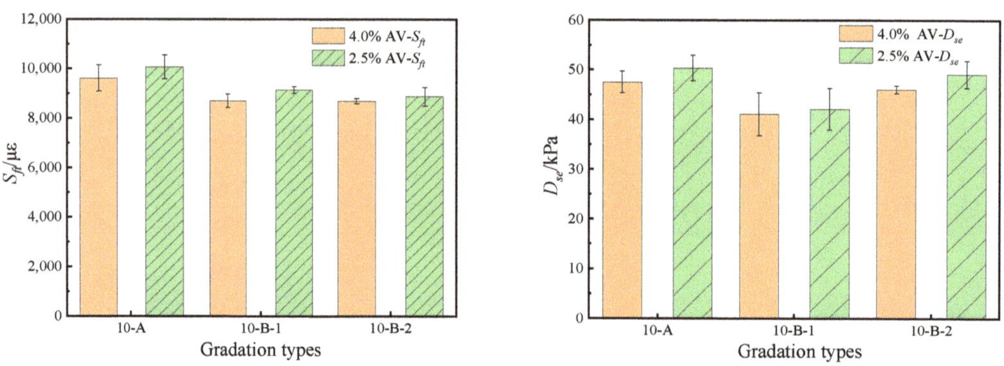

(a) **(b)**

Figure 7. Small beam bending test results of AR-SAMI @ 15 °C. (**a**) S_{ft}. (**b**) D_{se}.

(a) There were differences in the variation patterns between S_{ft} and D_{se} with air voids and gradation. Considering the different characterization meanings of the two indices, S_{ft} and D_{se}, should be selected for comprehensive evaluation of the crack resistance. Although S_{ft} and D_{se} at different temperatures showed the variation law of increasing with decreasing air voids, there are differences in the variation law of S_{ft} and D_{se} with different gradations (Figure 8). For example, for the 10-B-1 and 10-B-2 gradations, S_{ft} and D_{se} showed opposite patterns with the gradation change. According to the meaning, S_{ft} is the maximum strain when the specimen is damaged, indicating the deformation resistance of the mixture. However, this index is a single evaluation index, which cannot comprehensively evaluate the BBT results [38]. According to the principle of D_{se} calculation, the calculation of this index involves the whole process from loading to damage of the mixture, which can better reflect the comprehensive influence of multiple factors [40]. It can effectively evaluate the fracture resistance of the mixture under the test temperature conditions. Therefore, S_{ft} and D_{se} were recommended to evaluate the crack resistance comprehensively.

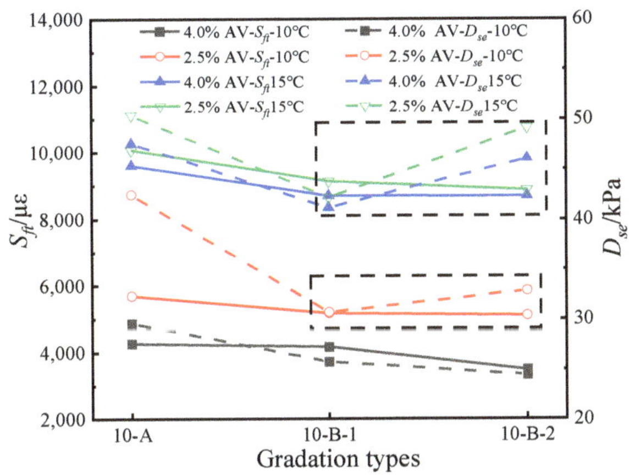

Figure 8. Change trend of S_{ft} and D_{se} with gradation.

(b) As the air voids decreased, the crack resistance of the mixture improved. The improvement of crack resistance indices was more significant at −10 °C when the air voids decreased to 2.5%. It was because when air voids were reduced, the internal compaction of the mixture was closer. Contact stress between the constituent particles was enhanced, which was conducive to resistance to external loads. Meanwhile, the asphalt film thickness increased at low air voids (Table 5), which strengthened the cohesion of the particles. This improved the resistance to deformation. The improvement of each index at different temperatures by reducing air voids is shown in Figure 8. The improvement range was only 2.1~6.7% at 15 °C. In comparison, it was 19.0~46.6% at −10 °C.

Table 5. Improvement of crack resistance by reducing air voids at different temperatures.

Gradation Types	Improvement of S_{ft}/%		Improvement of D_{se}/%	
	−10 °C	15 °C	−10 °C	15 °C
10-A	33.2	4.7	43.4	6.0
10-B-1	23.9	5.0	19.0	2.4
10-B-2	46.6	2.1	34.3	6.7

(c) The crack resistance of the 10-A gradation mixture was better than that of the other two groups. As shown in Figure 6, the S_{ft} and D_{se} values of the 10-A gradation mixture were higher than the other two groups at different air voids and test temperatures. According to the calculation results of the key sieve and Bailey method [37] parameters (as shown in Table 6), the 2.36 mm passing rate, 2.36–4.75 mm particle size content, and coarse aggregate ratio (*CA*) of 10-A gradation were significantly higher than the other two types. According to the calculation principle, the *CA* value indicates the proportion of 2.36–4.75 mm aggregate to 4.75–9.5 mm aggregate. Larger *CA* values indicate higher content of finer aggregates in the coarse aggregate. The higher fine aggregate content could effectively promote crack healing at the early loading stage. Thus, it might exhibit relatively high crack resistance performance.

Table 6. Evaluation parameters for each gradation.

Gradation Parameters		10-A	10-B-1	10-B-2
Passing rate of key sieve (2.36 mm)/%		40	28	29
Key sieve aggregate content/%	4.75–9.5 mm	34	60	60
	2.36–4.75 mm	17	7	9
Parameters of Bailey method	CA	0.40	0.11	0.15
	FA_c	0.48	0.61	0.52
	FA_f	0.47	0.59	0.53

Note: Referring to the calculation method of the Bailey method [37], CA is the coarse aggregate ratio; FA_c is the ratio of the coarse aggregate to the fine in the fine aggregate; FA_f is the ratio of the coarser aggregate to the finer aggregate in the finest aggregate.

3.2. Crack Extension Semicircular Bending Test

Each mixture was tested according to the CE-SCB test method mentioned above. The measured G_f and FI test results are shown in Figure 9.

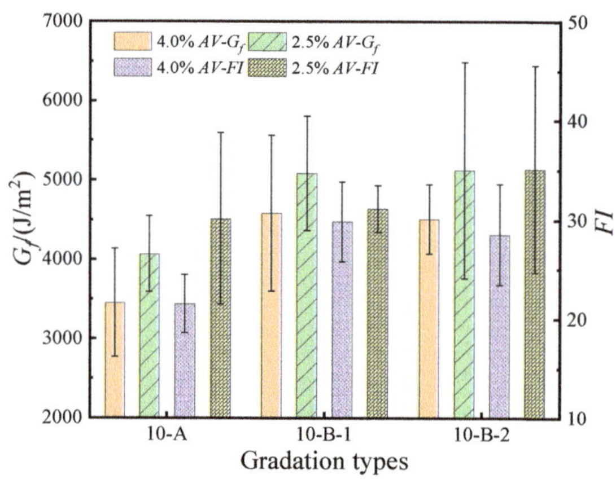

Figure 9. CE-SCB test results of AR-SAMI.

(a) The variation of G_f and FI were consistent with the change of air voids and gradation (Figure 9), indicating that both indices can be used to evaluate the crack expansion resistance of AR-SAMI. According to the meaning, G_f reflects the energy required for the crack expansion of the asphalt mixture. FI is the inherent property of the mixture material and characterizes the crack expansion resistance rate. The value of $|m|$ was small when FI was high. The slope of the force–displacement curve after the peak of the curve was flat indicates the better crack expansion resistance of AR-SAMI. In terms of the above test data, the two had good consistency.

(b) When the air voids were reduced, the crack extension resistance of AR-SAMI increased, and 10-A gradation had the maximum enhancement. With the reduction of air voids, the G_f and FI of each mixture increased, and the two indices increased by 17.8%, 11.0%, and 13.6%, and 40.0%, 4.4%, and 23.2%, respectively, indicating that reducing the air voids could enhance the crack expansion resistance of AR-SAMI. The mixture was compact due to the reduction of air voids. The decrease in inter-particle voids effectively inhibited the rapid development of cracks. It was demonstrated by the need for higher fracture energy to reach a fully extended crack. In addition, the two indices of the 10-A gradation mixture increased significantly higher than that of 10-B, indicating that the crack expansion resistance of AR-SAMI with more fine aggregate was more sensitive to the change of air voids.

(c) Among the three gradations, the crack expansion resistance of the 10-B type (including 10-B-1 and 10-B-2) mixture was significantly better than 10-A. As shown in Figure 9, the G_f and *FI* of 10-A were significantly lower than those of 10-B types. The indices of the 10-A at 2.5% air voids were still lower than the 10-B types with 4.0% air voids. It was because the passing rate of the 2.36 mm key sieve and *CA* value of 10-B types were smaller, indicating they had less fine aggregate content and coarse aggregate content of the finer aggregate. In contrast, the coarser aggregate was relatively more. In the presence of cracks, the more coarse aggregate facilitated the strain release at the cracks, which exhibited better resistance to crack expansion.

3.3. Low-Temperature Semicircular Bending Test

Each mixture was tested according to the LT-SCB test method mentioned above. G_f and K_{IC} results are shown in Figure 10.

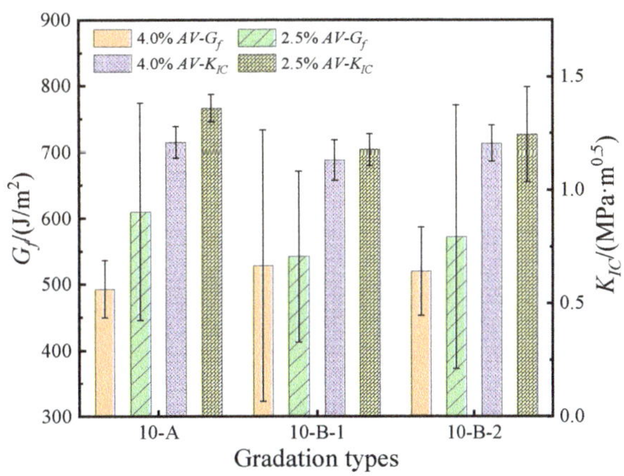

Figure 10. LT-SCB test results of AR-SAMI.

(a) The variation of G_f and K_{IC} was consistent with the change of air voids, while there were some differences with gradations. Considering the different focus of the two indices, G_f and K_{IC} were recommended to characterize the low-temperature crack expansion resistance of the AR-SAMI. As the air voids decreased, both indices increased. However, there was no significant correspondence with the change of gradation. For example, when the air voids were 4.0%, the G_f of the 10-A was the lowest, but its K_{IC} was the highest. The G_f reflects the energy required to resist deformation. It takes into account the whole cracking process. In comparison, K_{IC} demonstrates the ability to absorb energy in the fracture process. It takes into account the impact of the specimen load and size. So, there are differences between G_f and K_{IC}.

(b) G_f and K_{IC} increased as the air voids decreased. The 10-A gradation increased most significantly. When the design air voids were reduced from 4.0% to 2.5%, G_f and K_{IC} of three gradations increased. There were 23.7%, 2.7%, and 10.0% and 12.5%, 4.2%, and 3.2%, respectively. It showed that the mixture compactness increased with the air voids decreased. It was beneficial to improve the crack expansion resistance of AR-SAMI. Meanwhile, for the finer gradation (10-A), reducing the air voids could enhance the low-temperature crack resistance.

(c) The coupling influence of air voids complicated the influence of gradation on LT-SCB test results. When the air voids were 4%, G_f of 10-A was the lowest, and 10-B-1 was the highest. The indices showed an opposite magnitude relationship when the air voids decreased to 2.5%. G_f of 10-A was the highest and 10-B-1 was the lowest. It showed that the coarse aggregate content was higher at larger air voids (4%), thus with higher G_f. However, the improvement of G_f was more significant with the gradation with more fine aggregate after decreasing the air voids to 2.5%. In addition, the *FI* varied with gradation at different air voids in the same pattern, all of which were 10-A maximum.

3.4. Overlay Test

OT test was conducted at 25 °C with 1200 loadings. Then, *LLR* and *CRI* were calculated, respectively. The test results are shown in Figure 11.

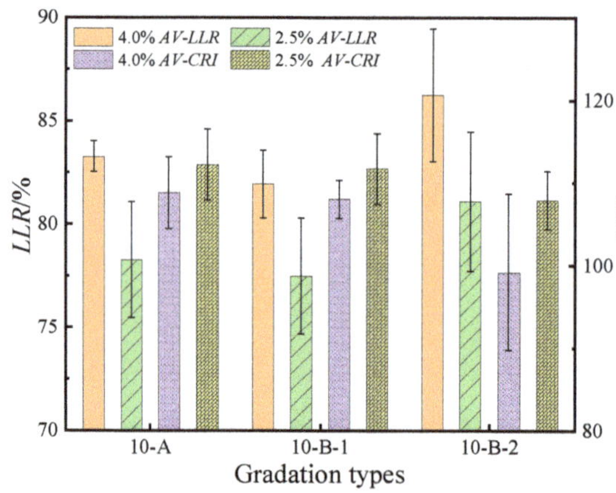

Figure 11. OT results of AR-SAMI.

(a) The *LLR* and *CRI* showed good consistency with the change of air voids and gradation. As air voids decreased, the *LLR* of each specimen decreased, and the *CRI* increased. Among the three gradations, the index values of 10-A and 10-B-1 were the same and better than that of 10-B-2. According to the meaning, the *LLR* indicates the load reduction ratio at cyclic tensile with fixed displacement. The *CRI* indicates the decreasing rate of loading. Although the evaluation of the two indices had differences, as far as the data are concerned, both indices could show the crack resistance of the AR-SAMI under the loading process.

(b) As the air voids decreased, the *LLR* decreased, and the *CRI* increased, indicating that the fatigue cracking resistance of the AR-SAMI was enhanced. The 10-B-2 test results were the worst among three gradations at two air voids. In addition, the *LLR* of all specimens was larger than 93% after 1200 cycles of loading, indicating that the AR-SAMI met the test requirements of the OT.

3.5. Correlation Analysis of Evaluation Indices

Among the above crack resistance evaluation tests and indices, there were differences in the evaluation focus. The Pearson Test was conducted using statistical analysis to further investigate the correlation between the above test methods and indices. Analysis results are shown in Figure 12.

(a) Among the five test methods, there were differences in the evaluation consistency of the respective two indices for each test. Among them, there was a significant correlation between the different evaluation indices of three test methods (−10 °C BBT, CE-SCB test, and OT). A single index could be used for alternative evaluation of the crack resistance of AR-SAMI. In the remaining two test methods (15 °C BBT and the LT-SCB test), their two indices did not have a significant correlation, indicating that their two indices of test methods had large differences. There did not exist simple substitution or conversion relationships. Therefore, both indices of the above two tests should be used to evaluate AR-SAMI comprehensively.

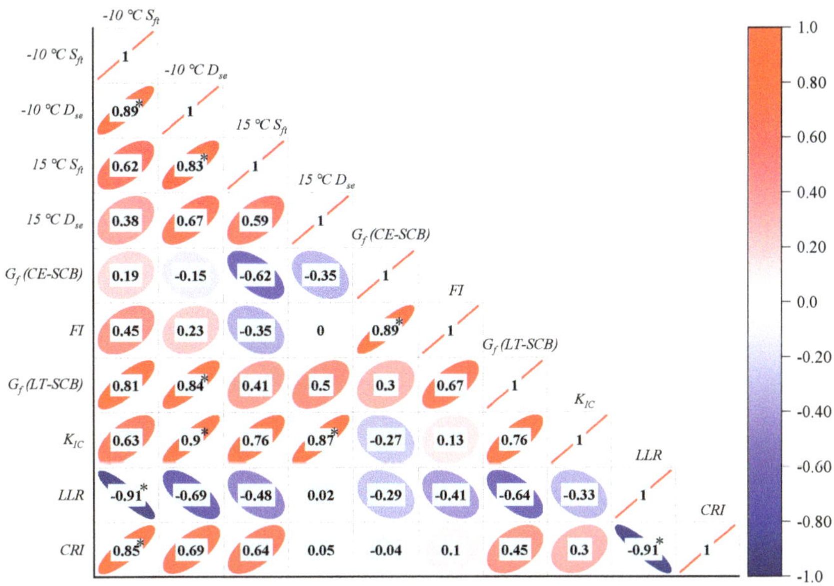

Figure 12. Pearson correlation analysis results for each evaluation index. Note: * indicates a significant correlation between the corresponding bivariate (significance parameter $p < 0.05$). The degree of elliptical flattening increases with the absolute value of the correlation coefficient. Red indicates positive correlation and blue indicates negative correlation. The brighter the color, the greater the correlation.

(b) Among the five test methods, there were good correlations between the indices of some test methods, and mutual substitution could be considered in the corresponding performance evaluation. The specific analyses are as follows:

- Evaluation of crack resistance: The two indices of the LT-SCB test were significantly correlated with the D_{se} of −10 °C BBT. So, the D_{se} of −10 °C BBT could be a substitute to evaluate the results of the LT-SCB test. However, indices of the LT-SCB test were not significantly related to the S_{ft} of the −10 °C BBT. Considering only S_{ft} was used in China [31], it is necessary to add D_{se} to evaluate crack resistance. In addition, studies have shown that asphalt mixtures are more likely to form cracks at low temperatures [33]. Therefore, instead of 15 °C BBT, −10 °C BBT results can be used to evaluate crack resistance.
- Evaluation of crack expansion resistance: Indices of the CE-SCB test were not significantly related to other test indices. However, there was a significant correlation between the two indices of this test. Since *FI* considered the post-peak load characteristics, it could better reflect the resistance of crack expansion. Therefore, the *FI* could be used to evaluate the crack expansion resistance of AR-SAMI.
- Evaluation of fatigue crack resistance: There was a significant correlation between the two indicators in the OT. The two OT indices were significantly correlated with the −10 °C S_{ft}. However, considering the significant differences between the two test conditions and crack types, a more general correlation still needs further verification. Therefore, using *CRI* to evaluate the fatigue crack resistance performance of AR-SAMI is still recommended.

Considering the validity and comprehensiveness of the evaluation indices and the simplicity of the test methods and indices, the following test methods and indices (as shown in Table 7) are recommended for the comprehensive evaluation of the crack resistance of AR-SAMI.

Table 7. Recommended crack resistance evaluation tests and indices.

Recommended Test Methods	Recommended Indices	Focus on Evaluating Performance
−10 °C BBT	S_{ft}, D_{se}	Low-temperature crack resistance
CE-SCB	*FI*	Crack extension resistance
OT	*CRI*	Fatigue crack resistance

3.6. Comprehensive Comparative Analysis of Crack Resistance

The recommended test methods and indices were adopted. Test results of crack resistance at different gradations and air voids were processed using the Min-Max normalization method. The normalized heat map of crack resistance is shown in Figure 13.

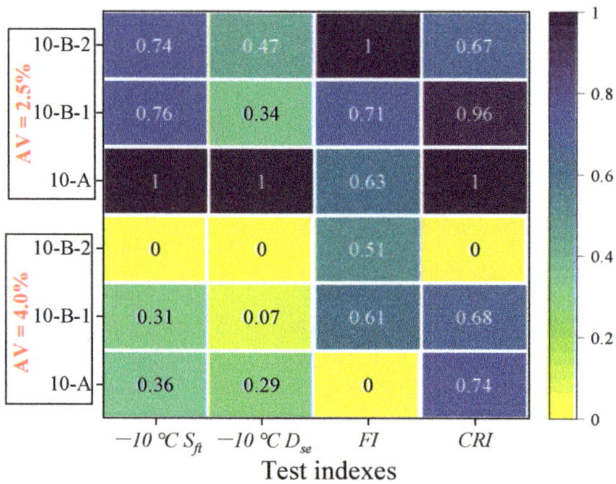

Figure 13. Crack resistance normalized heat map of AR-SAMI.

In summary, the crack resistance indices of the three gradations were better when the air voids were small. But, there were differences in the performance indices of different gradations. Comparing the four indices of the ARSAMI under two air voids, the indices of all gradations with 2.5% air voids were significantly higher than those of 4.0% air voids. Among all gradations, 10-A had the three highest indices (−10 °C S_{ft}, −10 °C D_{se}, and *CRI*) at two air voids. However, the *FI* of 10-A was the worst. It indicates that the crack resistance of AR-SMAI varies in different evaluation focus and even existed the opposite phenomenon.

Different test methods focus differently on evaluating crack resistance. The BBT is a single damage test that focuses on evaluating the strength characteristics of the AR-SAMI. The two SCB tests focus on evaluating crack expansion resistance under pre-crack conditions. The LT-SCB test focuses on low-temperature resistance to crack expansion performance. The OT focuses on performance decay under cyclic loading. It is similar to the fatigue performance test. Therefore, the design of the AR-SAMI needs to consider the evaluation results of the crack resistance under different tests. AR-SAMI should be a comprehensive balance design. For example, the BBT of AR-SAMI should be adopted for the new pavement construction because the sublayer cracks have not been formed. Suppose the AR-SAMI is added to the existing pavement. The CE-SCB test should be considered as the cracks already exist on the existing pavement with potential problems. Considering the OT can evaluate the tensile fatigue crack resistance performance, the OT should be a priority selection if the traffic volume is high. In addition, the inconsistency of evaluation results further indicated the necessity of using different test methods to comprehensively evaluate the crack resistance performance in various aspects of performance requirements.

4. Conclusions

AR-SAMI with two air voids and three gradations was selected. The BBT at different temperatures and CE-SCB, LT-SCB, and OT tests were conducted to evaluate the crack resistance. Statistical analysis of the test indices was performed. The main findings are as follows:

- All test methods showed that the crack resistance indices of AR-SAMI were significantly improved when the air voids were reduced from 4.0% to 2.5%. Therefore, the comprehensive crack resistance of AR-SAMI may perform better when the design air voids are decreased.
- The gradation 10-A, which contains more fine aggregate, performed better on most crack resistance indices. However, its crack extension resistance was poor. The CE-SCB and the LT-SCB tests showed that the 10-B type (including 10-B-1 and 10-B-2) with more coarse aggregate had better crack extension resistance under pre-existing cracks.
- There are differences in the evaluation results of different test methods and indices due to the differences in the evaluation mechanism. The test evaluation methods and corresponding indices should be selected according to the engineering characteristics of applications. The crack resistance should be designed in a targeted and comprehensive balance method. Considering the validity, comprehensiveness, and simplicity of the evaluation tests and indices, the test methods and indices in Table 7 were recommended for the comprehensive evaluation of the crack resistance of AR-SAMI.

In this work, the conclusions are based on three ten-type gradations. The type and number of test gradations are small, which has some limitations. Therefore, other mixture types and a wider range of gradations still need to be thoroughly analyzed for more universal research conclusions. In addition, the pavement performances (including high-temperature performance, low temperature performance, and water stability) and bonding properties of AR-SAMI should be combined for a comprehensive evaluation of the engineering application.

Author Contributions: Conceptualization, P.L. and Z.L.; methodology, W.P.; validation, S.T. and J.L.; formal analysis, W.P. and J.L.; investigation, J.L. and S.L.; resources, P.L. and Z.L.; data curation, W.P. and J.L.; writing—original draft preparation, S.T. and J.L.; writing—review and editing, W.P. and P.L.; supervision, P.L.; project administration, P.L.; funding acquisition, P.L. All authors have read and agreed to the published version of the manuscript.

Funding: This research was funded by the National Key Research and Development Program (China, 2021YFB2601000), Guangxi Key Research and Development Program (AB21220070, 2021AB22145), and the Postgraduate Scientific Research Innovation Project of Hunan Province (QL20220189).

Institutional Review Board Statement: Not applicable.

Informed Consent Statement: Not applicable.

Data Availability Statement: All data, models, and code generated or used during the study appear in the published article.

Conflicts of Interest: The authors declare no conflict of interest.

References

1. Wang, X.; Zhong, Y. Reflective crack in semi-rigid base asphalt pavement under temperature-traffic coupled dynamics using XFEM. *Constr. Build. Mater.* **2019**, *214*, 280–289. [CrossRef]
2. Oshone, M.; Dave, E.V.; Sias, J.E. Asphalt mix fracture energy based reflective cracking performance criteria for overlay mix selection and design for pavements in cold climates. *Constr. Build. Mater.* **2019**, *211*, 1025–1033. [CrossRef]
3. Idris, I.I.; Sadek, H.; Hassan, M. State-of-the-Art Review of the Evaluation of Asphalt Mixtures' Resistance to Reflective Cracking in Laboratory. *J. Mater. Civil Eng.* **2020**, *32*, 0312004. [CrossRef]
4. Rith, M.; Lee, S.W. Development of cohesive-zone-based prediction model for reflective cracking in asphalt overlay. *Int. J. Pavement Eng.* **2022**, *23*, 1050–1059. [CrossRef]
5. Shan, J.; Zhang, Y.; Wu, S.; Wu, Q.; Jiao, Y.; Du, Y. Cracking behavior of asphalt pavement with a graded gravel layer based on computational granular mechanics. *Constr. Build. Mater.* **2022**, *345*, 128199. [CrossRef]

6. Fang, H.; Luo, H.; Zhu, H. The feasibility of continuous construction of the base and asphalt layers of asphalt pavement to solve the problem of reflective cracks. *Constr. Build. Mater.* **2016**, *119*, 80–88. [CrossRef]
7. Shafabakhsh, G.; Ahmadi, S. Reflective cracking reduction by a comparison between modifying asphalt overlay and sand asphalt interlayer: An experimental evaluation. *Int. J. Pavement Eng.* **2021**, *22*, 192–200. [CrossRef]
8. Baek, C. Performance Evaluation of Fiber-Reinforced, Stress Relief Asphalt Layers to Suppress Reflective Cracks. *Appl. Sci.* **2020**, *10*, 7701. [CrossRef]
9. Haslett, K.; Dave, E.; Sias, J.; Linder, E. Statistical Analysis Framework to Evaluate Asphalt Concrete Overlay Reflective Cracking Performance. *Transp. Res. Rec. J. Transp. Res. Board* **2022**, *2676*, 132–146. [CrossRef]
10. Habbouche, J.; Hajj, E.Y.; Morian, N.E.; Sebaaly, P.E.; Piratheepan, M. Reflective cracking relief interlayer for asphalt pavement rehabilitation: From development to demonstration. *Road Mater. Pavement* **2017**, *18*, 30–57. [CrossRef]
11. Pasquini, E.; Pasetto, M.; Canestrari, F. Geocomposites against reflective cracking in asphalt pavements: Laboratory simulation of a field application. *Road Mater. Pavement* **2015**, *16*, 815–835. [CrossRef]
12. Zeng, Z.; Guo, Y.; Li, P.; Shen, A.; Zhai, C. Performance research of fiber-reinforced asphalt rubber as a stress-absorbing membrane interlayer. *J. Adhes. Sci. Technol.* **2021**, *35*, 2047–2063. [CrossRef]
13. Zhang, K.; Zhang, Z.; Luo, Y. Material Composition Design and Anticracking Performance Evaluation of Asphalt Rubber Stress-Absorbing Membrane Interlayer (AR-SAMI). *Adv. Mater. Sci. Eng.* **2018**, *2018*, 8560604. [CrossRef]
14. Ogundipe, O.M.; Thom, N.H.; Collop, A.C. Evaluation of performance of stress-absorbing membrane interlayer (SAMI) using accelerated pavement testing. *Int. J. Pavement Eng.* **2013**, *14*, 569–578. [CrossRef]
15. Karki, P.; Mraiza, Z.; Karnei, E.; Hu, S. Performance-graded asphalt binder selection catalog for asphalt overlays. *Constr. Build. Mater.* **2022**, *319*, 126012. [CrossRef]
16. Peng, W.; Li, P.; Gong, W.; Tian, S.; Wang, Z.; Liu, S.; Liu, Z. Preparation and mechanism of rubber-plastic alloy crumb rubber modified asphalt with low viscosity and stabilized performance. *Constr. Build. Mater.* **2023**, *388*, 131687. [CrossRef]
17. Picado-Santos, L.G.; Capitão, S.D.; Neves, J.M.C. Crumb rubber asphalt mixtures: A literature review. *Constr. Build. Mater.* **2020**, *247*, 118577. [CrossRef]
18. Asgharzadeh, S.M.; Sadeghi, J.; Peivast, P.; Pedram, M. Fatigue properties of crumb rubber asphalt mixtures used in railways. *Constr. Build. Mater.* **2018**, *184*, 248–257. [CrossRef]
19. Aliha, M.R.M.; Fazaeli, H.; Aghajani, S.; Moghadas Nejad, F. Effect of temperature and air void on mixed mode fracture toughness of modified asphalt mixtures. *Constr. Build. Mater.* **2015**, *95*, 545–555. [CrossRef]
20. Zhang, Q. Effect of Grading Type on the Performance of Warm-Mix Rubber-Asphalt Mixture. *IOP Conf. Ser. Earth Environ. Sci.* **2020**, *526*, 12152. [CrossRef]
21. Pan, R.; Li, Y. Effect of warm mix rubber modified asphalt mixture as stress absorbing layer on anti-crack performance in cold region. *Constr. Build. Mater.* **2020**, *251*, 118985. [CrossRef]
22. Wang, S.; Yan, K.; Ge, D.; Hong, Z. Laboratory research on the performance of stress-absorption interlayer (SAI) of waste tire rubber and amorphous ploy alpha olefin modified asphalt. *Constr. Build. Mater.* **2019**, *223*, 830–840. [CrossRef]
23. Thives, L.P.; Pais, J.C.; Pereira, P.A.A.; Minhoto, M.C.; Trichês, G. Performance of Asphalt Rubber Mixture Overlays to Mitigate Reflective Cracking. *Materials* **2022**, *15*, 2375. [CrossRef] [PubMed]
24. Pei, Z.; Yi, J.; Mao, Q.; Feng, D.; Wang, D. DEM analysis of the evolution of reflection cracks in old cement concrete pavement with an ATB layer. *Int. J. Pavement Eng.* **2022**, 1–10. [CrossRef]
25. Li, C.; Zhou, H.; Wu, Z.; Wang, Z. International Conference on Transportation Engineering. In *Alleviating Crack Properties of Crack Alleviating Mixture*; American Society of Civil Engineers: Reston, VA, USA, 2009.
26. *AASHTO TP 124-16*; American Association of State Highway and Transportation Officials Standard Method of Test for Determining the Fracture Potential of Asphalt Mixtures Using Semicircular Bend Geometry (SCB) at Intermediate Temperature. American Association of State Highway and Transportation Officials: Washington, DC, USA, 2016.
27. *AASHTO TP 105-13*; American Association of State Highway and Transportation Officials Standard Method of Test for Determining the Fracture Energy of Asphalt Mixtures Using the Semicircular Bend Geometry (SCB). American Association of State Highway and Transportation Officials: Washington, DC, USA, 2019; p. 13.
28. Germann, F.P.; Lytton, R.L. *Methodology for Predicting the Reflection Cracking Life of Asphalt Concrete Overlays*; Texas A&M University: College Station, TX, USA, 1979; pp. 25–45.
29. Nam, B.H.; Golestani, B.; Noori, M.; Tatari, O.; An, J. *Investigation of Reflective Cracking Mitigation Techniques*; Department of Civil, Environmental, and Construction Engineering University of Central Florida: Orlando, FL, USA, 2014; pp. 1–153.
30. Zhou, F.; Newcomb, D.; Gurganus, C.; Banihashemrad, S. *Field Validation of Laboratory Tests to Assess Cracking Resistance of Asphalt Mixtures: An Experimental Design (2016)*; National Academy of Sciences: Washington, DC, USA, 2016.
31. *JTG F40-2004*; Ministry of Communications of the People's Republic of China Technical Specifications for Construction of Highway Asphalt Pavement. Ministry of Communications: Beijing, China, 2004.
32. Wang, H.; Dang, Z.; Li, L.; You, Z. Analysis on fatigue crack growth laws for crumb rubber modified (CRM) asphalt mixture. *Constr. Build. Mater.* **2013**, *47*, 1342–1349. [CrossRef]
33. Mandal, T.; Yin, H.; Ji, R. Correlating laboratory and full-scale reflective cracking tests for airfield pavements. *Constr. Build. Mater.* **2018**, *169*, 47–58. [CrossRef]

34. *JTG E20-2011*; Ministry of Communications of the People's Republic of China Standard Test Methods of Bitumen and Bituminous Mixtures for Highway Engineering. Ministry of Communications: Beijing, China, 2011.
35. *DB45/T 1098-2014*; Guangxi Zhuang Autonomous Region Quality and Technical Supervision Bureau Technical Specification for Construction of Asphalt Rubber Pavement. Guangxi Zhuang Autonomous Region Quality and Technical Supervision Bureau: Nanning, China, 2014.
36. *DG/TJ08-2109-2012*; Shanghai Urban-Rural Construction and Transportation Commission Technical Code for Asphalt Rubber Pavement. Shanghai Urban-Rural Construction and Transportation Commission: Shanghai, China, 2012.
37. Swathi, M.; Andiyappan, T.; Guduru, G.; Amarnatha Reddy, M.; Kuna, K.K. Design of asphalt mixes with steel slag aggregates using the Bailey method of gradation selection. *Constr. Build. Mater.* **2021**, *279*, 122426. [CrossRef]
38. Ge, Z.; Huang, X.; Xu, G. Evaluation of asphalt-mixture's low-temperature anti-cracking performance by curvature strain energy method. *J. Southeast Univ. Nat. Sci. Ed.* **2002**, *32*, 653–655.
39. *Tex-248-F*; Texas Department of Transportation Test Procedure for Overlay Test. Texas Department of Transportation: Austin, TX, USA, 2017; pp. 1–13.
40. Feng, D.; Cui, S.; Yi, J.; Chen, Z.; Qin, W. Evaluation Index of Low-temperature Asphalt Mixture Performance Based on Semi-circular Bending Test. *China J. Highw. Transp.* **2020**, *33*, 50–57.

sustainability

MDPI

Article

Effects of Rejuvenator Dosage, Temperature, RAP Content and Rejuvenation Process on the Road Performance of Recycled Asphalt Mixture

Tingting Jiang [1,*], Qiaojuan Fan [1], Mingye Hou [2], Shuzhen Mi [1] and Xiaohui Yan [3]

[1] Henan Vocational College of Water Conservancy and Environment, Zhengzhou 450008, China
[2] Henan Communications Planning & Design Institute Co., Ltd., Zhengzhou 450000, China
[3] Shanghai Urban Construction Vocational College, Shanghai 201415, China
* Correspondence: jtt872302967@163.com; Tel.: +86-18839785541

Abstract: In this paper, the key technologies in the construction process of hot in-place recycling were investigated in order to improve the utilization rate of waste asphalt mixture; traditional lab tests including penetration, softening point and ductility tests, atomic force microscope test of recycled asphalt under different rejuvenator content, and the test of milling on grading at different temperatures were carried out. The influence of RAP content and rejuvenation processes on road performance were studied, and the low-temperature performance of mixture was analyzed by the energy analysis method, and the evaluation index was proposed. Test results indicated that the penetration and ductility increases, the softening point decrease with the rejuvenator content increasing, and the optimum rejuvenator content is 4%. The optimum mixing and compaction temperature will decrease by 2–6 °C on average for every 10% increase of RAP content by analyzing the mixture volume index. The results showed enhance rutting resistance of the mixture but lower moisture resistance and low-temperature crack resistance by adding the RAP content. The strain energy density of 10 KJ/m^3 is proposed to evaluate the low-temperature performance of the mixture, and 30% RAP produces optimal mixture. The higher rutting resistance and moisture resistance can be obtained by using the construction process of RAP+ rejuvenator co-heating, and higher low-temperature crack resistance with RAP+ rejuvenator without heating.

Keywords: rejuvenator dosage; microstructure; RAP content; rejuvenation process; road performance

Citation: Jiang, T.; Fan, Q.; Hou, M.; Mi, S.; Yan, X. Effects of Rejuvenator Dosage, Temperature, RAP Content and Rejuvenation Process on the Road Performance of Recycled Asphalt Mixture. *Sustainability* **2023**, *15*, 3539. https://doi.org/10.3390/su15043539

Academic Editors: Dawei Wang, Chaohui Wang, Kai Liu and Qian Chen

Received: 27 December 2022
Revised: 12 January 2023
Accepted: 13 February 2023
Published: 14 February 2023

1. Introduction

In recent years, with the rapid development of highways, particularly of asphalt pavement, older highways have gradually entered the maintenance period. A large number of pavement maintenance and repair work has resulted in such problems as the increase of reclaimed asphalt pavement (RAP). Especially, a large amount of waste asphalt pavement materials, if not timely recycled, not only take up a lot of land and is a waste of resources, but also pollute the environment [1–3]. Studies have shown that the reuse of RAP was significant in saving resources, protecting the environment and achieving sustainable development of the construction industry [4–8].

Generally, the asphalt pavement recycling technology is divided into four types, namely hot central plant recycling, hot in-place recycling, cold central plant recycling and cold in-place recycling. Among them, the hot in-place recycling technology has the advantages of rapid construction, maximum recycling of RAP, saving transportation costs, and benefiting interlayer bonding, which is of great significance to the development of environment-friendly highway maintenance technology [9–13].

Recently, extensive studies focused on the performance characterization and microscopic mechanism of recycled asphalt mixtures with different types or dosage of rejuvenator. The results showed that the content of the regenerating agent, the type of regenerating agent and the production technology all had influence on the properties of asphalt mixture. Ding, L.T. et al. [14] investigated the morphology and property changes of the aged asphalt with a rejuvenator. The results indicated that a rejuvenator was conducive to microstructure change and high temperature rheological properties. Yu, J. et al. [15] comparatively analyzed the rejuvenator effect of artificial aged asphalt with soft bitumen, liquid surfactant and bio-rejuvenator. The results indicated that the bio-rejuvenator performed better with additional content of artificial RAP asphalt. Ma, Y.T. et al. [16] presented the adding procedure of a rejuvenator would influence rheological and aging characteristics of aged asphalt. Yaseen, G. et al., Chen, A.Q. et al. and Bilema, M. et al. [17–19] surveyed the influence of rejuvenator content on penetration, softening point and ductility of recycled asphalt. Researchers investigated the microscopic mechanism of aged asphalt with rejuvenator, and it was found that the composition of aged asphalt changed by the rejuvenator, the light components decreased and the micelle number and proportion in the colloidal structure increased [20,21]. Based on the results of these analyses, there is a consensus that a rejuvenator can effectively control and reduce the performance loss of recycled asphalt after aging and change the colloidal structure of aged asphalt [22].

In addition, it was observed that appropriate RAP content has a significant effect on the performance of hot recycled asphalt mixture [23–27]. Bilema, M. et al. [28] concluded that this enhanced stiffness and rutting resistance of the RAP but lowered moisture resistance, and the pavement sustainability and rutting performance of RAP could be restored by adding WFO and CR. Yin, P. et al. [29] found that with the addition of RAP, high-temperature performance of HRAM increased and the low-temperature performance and moisture susceptibility decreased. Zhu, J.Q. et al. [30] surveyed high modulus asphalt mixture containing RAP, and it was found that the dynamic modulus values and moisture damage susceptibility of asphalt mixture were optimal with 40% RAP. Based on the results of vehicle load simulator, Zaumanis, M. et al. [31] demonstrated that the crack resistance of recycled high modulus asphalt mixture was deficiency when RAP content was 100%. Ma, Y.T. et al. [32] revealed that mixtures heated with a higher temperature had better cracking and moisture resistance performances. Li, X.L. et al. [33] indicated that the heating temperature of RAP and mixing time could improve the homogeneity of RAP mixture. Furthermore, the rejuvenation process was taken into considered as key factor of performance of hot in-place recycling asphalt mixtures. Xie, Z.X. et al. [34] investigated the effect of rejuvenator types and mixing procedures on volumetric properties of mixtures. The results indicated that some volumetric properties of mixtures changed under different mixing procedures and rejuvenator types. Lei, Y. et al. and Lei, Y. et al. [35,36] investigated that the high temperature performance, compaction property and workability have obvious differences with different mixing sequences.

Overall, the rejuvenator, temperature, RAP content and rejuvenation process are all crucial factors to achieve the highest performance of the hot in-place recycling asphalt mixtures [37]. However, most studies mainly focused on the effect of the rejuvenator on aged asphalt and the effect of RAP content on aged asphalt mixture, lacking research on the performance of recycled asphalt mixture systematic research. In this paper, the optimum dosage of rejuvenator was studied first, and then the optimum mixing and compaction temperature was determined based on four kinds of hot recycled asphalt mixture with different RAP content. Furthermore, the influence of different RAP content and rejuvenation process on the performance of the mixture was discussed. The research results were of great significance to improve the construction quality of hot in-place recycling.

2. Materials and Methods

2.1. Materials

The materials in this study included asphalt extraction from old road milling materials, No. 90 virgin asphalt, one type of permeable rejuvenator, and reclaimed asphalt pavement (RAP). Aged asphalt and RAP were provided by a local old road in Henan, China. The permeable rejuvenator was LT-2, independently developed by a company in Wuhan, Hubei,

China. The test methods were carried out in accordance with Chinese standard JTG E20-2011 [38] and the properties of permeable rejuvenator are shown in Table 1.

Table 1. Properties of permeable rejuvenator.

Property Indices	Test Results	Test Methods
Viscosity at 60 °C (mm^2/s)	87	T0619-2011
Flash point (°C)	238	T0633-2011
Saturates (%)	7.3	T0618-1993
Aromatics (%)	69.5	T0618-1993
Viscosity ratios before and after RTFOT	2.3	T0610-2011 T0619-2011
Mass ratios before and after RTFOT	1.5	T0610-2011
15 °C Density (g/cm^3)	0.94	T0603-2011
Appearance	Black viscous liquid	

2.2. Testing Method

2.2.1. Optimum Dosage of Rejuvenator Test

1. Traditional laboratory tests of asphalt

In this paper, the performance of recycled asphalt mixed with different rejuvenator ratio (0, 3%, 4%, 5%) was evaluated by penetration, softening point and ductility tests. One group, two groups and three groups of recycled asphalt specimens were formed for penetration, softening point and ductility tests for each rejuvenator content, respectively, where three test points were selected on the penetration sample. The process of testing specimen was carried out in accordance with Chinese standard JTG E20-2011 [38].

2. Atomic force microscopy (AFM) test

The surface morphology of the recycled asphalt mixed with different rejuvenator ratio (0, 3%, 4%, 5%) was observed by NanoMan vs. AFM. The specimen was tested at a temperature of 25 °C and a relative humidity of 25%. The micro images in the range of 20 um × 20 um were presented by tapping mode at no less than 3 positions on each sample surface. Further, the data such as surface roughness were calculated by NanoScope Analysis software. The calculation formulas are:

$$R_{max} = h_{max} - h_{min} \tag{1}$$

$$R_q = \sqrt{\frac{\sum_1^N Z_i^2}{N}} \tag{2}$$

$$R_a = \frac{1}{N}\sum_{j=1}^{N}|Z_j| \tag{3}$$

where R_{max} is the maximum vertical distance between a highest and a lower image data point in an image processed. R_q and R_a represent the roughness of the image, R_q is the root mean square average of the adhesion deviation of the image data plane, R_a represents the arithmetic average of the absolute value of the surface adhesion deviation measured from the average plane, and n is defined as the number of 'bee structures' in the image range. h_{max} and h_{min} are the maximum and minimum height in the atomic force microscope image, respectively, and Z_i and Z_j represent adhesion deviation.

2.2.2. Temperature Control Test

1. Milling temperature control test

The environmental temperature, wind speed and solar radiation of the test site were tested by the automatic environmental monitoring system, and the surface temperature were tested by infrared thermometer. The relationship between various factors and heating temperature was analyzed, and the influence of temperature on the hot in-place rejuvenation was obtained. The gradation design of milling hot recycled asphalt mixture was carried out at 120 °C, 130 °C and 140 °C, and the influence of milling on the gradation of recycled asphalt mixture at different temperatures was analyzed.

2. Optimum mixing and compaction temperature test

In this paper, the mixture gradations with each RAP content are the same mineral aggregate as the Standard median gradation. The optimum asphalt–aggregate ratio of each mixture was determined by the Marshall design method according to the Chinese standard JTG E20-2011 [38], which was 3.95%, 3.67%, 3.45% and 3.32% for the mixture with RAP content (30%, 40%, 50%, and 60%), respectively. For each RAP content, five groups of Marshall specimens were formed with 165 °C as the median change of mixing temperature and 145 °C as the median change of compaction temperature, respectively, with four parallel specimens in each group. The Marshall test was conducted on hot recycled asphalt mixture with RAP content of 30%, 40%, 50% and 60% by using the volume index control method. The Marshall test was conducted on hot recycled asphalt mixture with RAP content of 30%, 40%, 50% and 60% by using the volume index control method. The optimum mixing and compaction temperature was when the air void content was 4%.

2.2.3. Road Performance Test

In this paper, the effects of RAP content (30%, 40%, 50%, 60%) on high-temperature stability, moisture-induced damage and low-temperature crack resistance of hot recycled asphalt mixture were evaluated by rutting test, immersion Marshall test, freeze-thaw split test and low-temperature bending test. However, only the flexural tensile strength or flexural tensile strain determined in the low-temperature bending test was used as a comprehensive evaluation of the performance of recycled asphalt mixture under low-temperature conditions, which can only replace its unilateral flexural tensile strength. Therefore, the strain energy density was introduced from the perspective of energy as a comprehensive consideration to evaluate the performance index of recycled asphalt mixture under low-temperature conditions. The calculation formula is shown as Equation (4).

$$W_f = \frac{dW}{dV} = \int_0^{\varepsilon_0} \sigma_{ij} d\varepsilon_{ij} \tag{4}$$

where W_f is the strain energy density, σ_{ij} and ε_{ij} are the stress and strain components, ε_0 is the strain critical value.

2.2.4. Effect of Rejuvenation Process on Road Performance of Recycled Asphalt Mixture

In this paper, the recycled asphalt mixture with 30% RAP content were molded according to three different rejuvenation construction processes under the Optimum mixing and compaction temperature, and the road performance of the recycled asphalt mixture with different mixing processes were evaluated by the rutting test, freeze-thaw split test and low-temperature bending test. The different rejuvenation construction processes are shown in Table 2.

Table 2. Different rejuvenation construction processes.

Number	Rejuvenation Construction Process
Process I	RAP Heating + Rejuvenator
Process II	RAP + Rejuvenator Co-heating
Process III	RAP +Rejuvenator not heating

3. Results and Discussion

3.1. Effect of Optimum Dosage of Rejuvenator

3.1.1. Rejuvenator Content

Table 3 show the penetration, softening point and ductility of different samples obtained from three index tests as a function of rejuvenator dosage. From Table 3, it can be concluded that the penetration and ductility of the recycled asphalt increases and the softening point decrease with increasing rejuvenator dosage. During the production process of long-term repair and application of asphalt binder, slight component loss, relative decrease of colloidal element content and significant increase of asphaltene content will occur, which will cause the increase of asphalt softening point and decrease of penetration and ductility, and the permeable rejuvenator has a protective effect.

Table 3. Asphalt performance of different rejuvenator dosage.

Rejuvenator Dosage (%)	Penetration (25 °C, 0.01 mm)	Softening Point (°C)	Ductility (10 °C, cm)	Remark
0	24.6	70	8.7	
3	30.8	66.3	12.2	Permeable Rejuvenator
4	35.5	64.2	15.6	
5	41.2	60.9	18.7	

3.1.2. Nano-Morphology of Asphalt Sample

The two-dimensional height scanning results of the samples are shown in Figure 1a–e. From Figure 1, it can be clearly seen that the asphalt samples with different rejuvenator dosage have a 'bee structure', and the microstructure of aged asphalt and the nanomorphology in the AFM image have also changed to varying degrees with the increase of rejuvenator dosage. Compared with the virgin asphalt, the aging asphalt 'bee structure' agglomeration phenomenon was more serious, the size was larger, the number was less, and as a whole showed a wide and rare phenomenon. With the increase of the rejuvenator dosage, the size of 'bee structure' decreased and the number increased, which tended to the microstructure of virgin asphalt, while there was still a small amount of large 'bee structures'. During asphalt aging, the light components in asphalt would gradually decrease, and the proportion of asphaltene in asphalt would gradually increase. The existence of asphaltene was the main reason for the formation of 'bee structure', so the 'bee structure' of aging asphalt was more obvious than that of virgin asphalt. The composition of the aged asphalt was changed by the rejuvenator, which also reduced the relative content of the aged asphaltene, so that the original dense 'bee structure' development was more loose and smaller in size, closer to the virgin asphalt.

Figure 2a–e shows the fluctuation trend of 'bee structure' of asphalt samples with different rejuvenator dosage from a three−dimensional perspective. From Figure 2, it can be concluded that with the increase of the rejuvenator dosage, the coarse columnar structure of the aged asphalt gradually became finer and more abundant, and the surface roughness was obviously improved, which was closer to the virgin asphalt, indicating that the aged asphalt had a better rejuvenation effect.

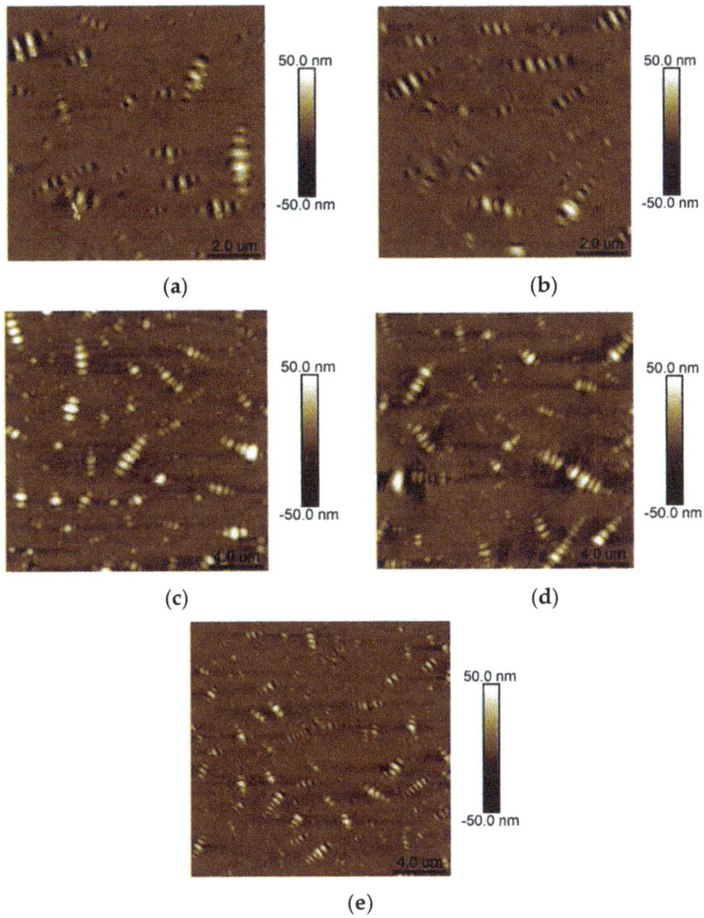

Figure 1. Two−dimensional height diagram of asphalt samples with different rejuvenator dosage: (**a**) Aged asphalt; (**b**) Asphalt mixed with 3% rejuvenator; (**c**) Asphalt mixed with 4% rejuvenator; (**d**) Asphalt mixed with 5% rejuvenator; (**e**) Virgin asphalt.

The number, R_{max}, roughness and other parameters of 'bee structure' of asphalt samples with different rejuvenator dosage were analyzed, and the average value of each test result was taken. The statistical results are shown in Table 4. It is concluded from Table 4 that with the increase of the rejuvenator dosage, roughness indexes R_q and R_a of recycled asphalt gradually decreased, and R_{max} decreased first and then increased. Based on the changes of R_q, R_a and R_{max}, it can be seen that the R_{max} of recycled asphalt mixed with 4% rejuvenator was closer to that of virgin asphalt, indicating that the addition of recycled agent had caused a certain rejuvenation of aged asphalt. Combined with the three index test results, it can be obtained that the optimum content of rejuvenator was 4%.

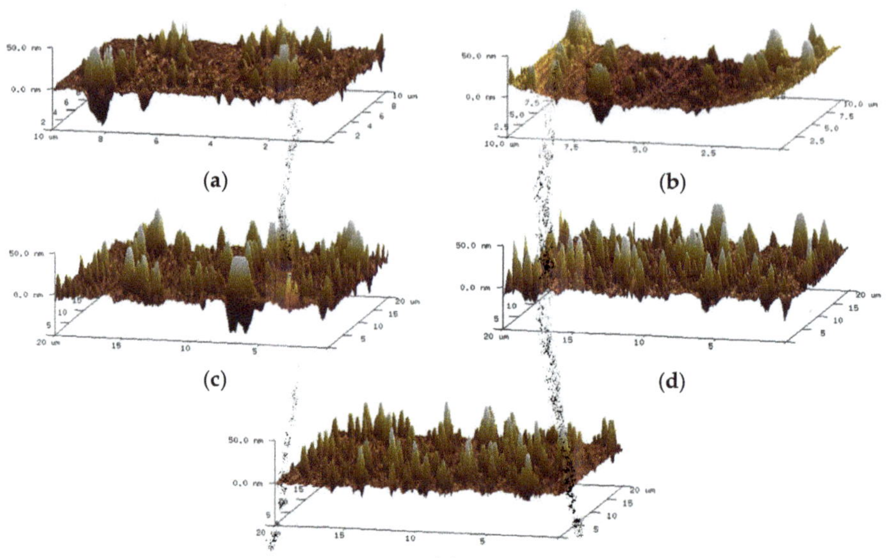

Figure 2. Three−dimensional height diagram of asphalt samples with different rejuvenator dosage: (**a**) Aged asphalt; (**b**) Asphalt mixed with 3% rejuvenator; (**c**) Asphalt mixed with 4% rejuvenator; (**d**) Asphalt mixed with 5% rejuvenator; (**e**) Virgin asphalt.

Table 4. The results of parameter analysis.

Sample Shape	Aged Asphalt	Asphalt Mixed with			Virgin Asphalt
		3% Rejuvenator	4% Rejuvenator	5% Rejuvenator	
n	34	38	76	59	79
R_{max}/nm	118	108	98.6	115	82.1
R_q/nm	9.42	9.5	5.04	8.15	5.48
R_a/nm	5.68	7.58	2.68	4.82	2.93

3.2. *Effect of Temperature Control*

3.2.1. Milling Temperature Control

The relationship between ambient temperature, wind speed, solar radiation and surface temperature of structure layer is shown in Figure 3a–c. It can be concluded from Figure 3 that the ambient temperature, wind speed and solar radiation had a significant effect on the heating temperature. The surface temperature of the structural layer increased significantly with the increase of ambient temperature and solar radiation, and decreased with the increase of wind speed.

It is necessary to control a reasonable heating temperature during construction, so that a certain temperature gradient is formed on the road surface and the temperature can be effectively transferred for a long enough time, as shown in Figure 4.

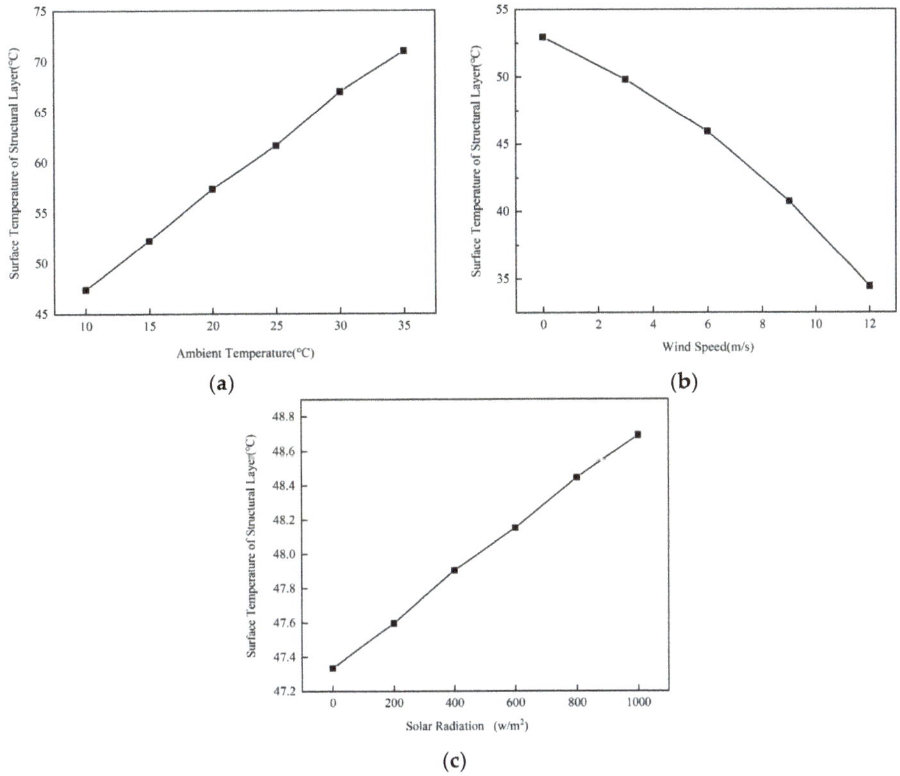

Figure 3. Environment–temperature diagram: (**a**) Effect of ambient temperature on surface temperature of structural layer; (**b**) Effect of wind speed on surface temperature of structural layer; (**c**) Effect of solar radiation on surface temperature of structural layer.

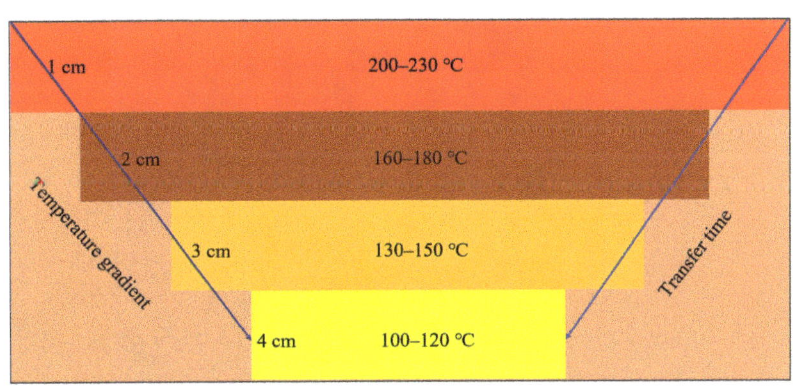

Figure 4. Temperature gradient and time diagram.

The influence of milling on gradation at different temperatures is shown in Figure 5. It can be concluded from Figure 5 that when the pavement was milled under different temperature conditions, the gradation of the pavement after milling changed greatly. The lower the milling temperature, the more the gradation after milling deviates from the upper limit of gradation. The higher the milling temperature, the closer the gradation after milling was to the original gradation.

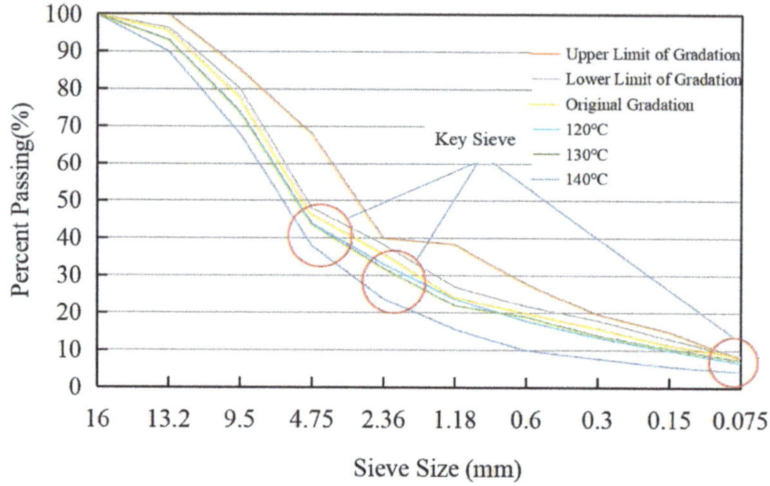

Figure 5. Effect of Different Temperature Milling on Gradation.

Figure 6 shows the effect of pavement milling at 120 °C. The temperature interferes in the breakdown of the old asphalt layer and in the dimensions of the "black aggregates" obtained. At low temperature, the aggregate milling crushing of asphalt pavement increases, and the aging of old asphalt decreases. At high temperature, it can effectively reduce the damage of milling to aggregate and increase the rate of old asphalt recovered, effectively ensure the uniformity of the grading of the original pavement.

Figure 6. Illustration of surface milling at low temperature.

3.2.2. Optimum Mixing and Compaction Temperature

The change of air void content of Marshall specimens under different RAP content, mixing and compaction temperature is shown in Figure 7a,b. The optimum mixing and compaction temperature test results of hot recycled asphalt mixture is shown in Figure 8. From Figures 7 and 8, it can be concluded that optimum mixing and compaction temperature of hot recycled asphalt mixtures decreased with RAP content, when the content of RAP was 30%, 40%, 50% and 60%, the optimum mixing temperature was 175 °C, 170 °C, 165 °C and 155 °C, and the optimum compaction temperature was 152 °C, 150 °C, 144 °C and 139 °C; for every 10% increase in RAP content, the mixing and compaction temperature decreased by 2–6 °C.

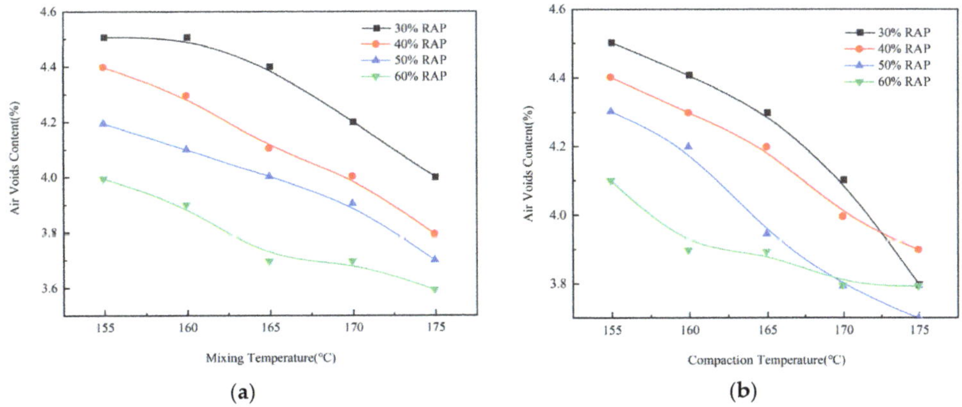

Figure 7. Air void content changes at different mixing and compaction temperatures: (**a**) Relationship between mixing temperature and air void content; (**b**) Relationship between compaction temperature and air void content.

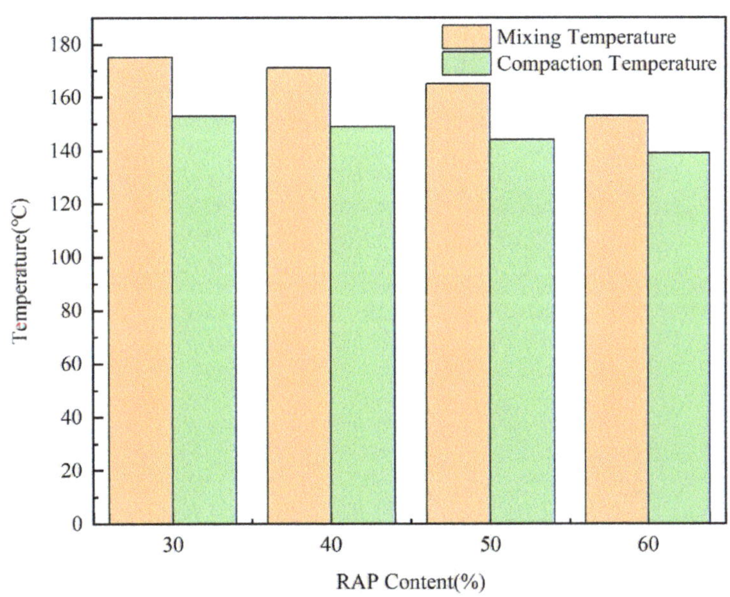

Figure 8. The optimum mixing and compaction temperature under different RAP content.

The possible reason was that asphalt film has a remarkable bonding effect when high RAP is added to the hot in-place recycling technology, the aged asphalt in RAP was well bonded with the old aggregate, and the recycled asphalt formed after adding the rejuvenator and virgin asphalt was easy to be wrapped with the old aggregate and pressed into a dense structure; with the increase of RAP content, recycled asphalt needs lower heat to soften itself to improve fluidity under the bonding effect of asphalt film, so as to bond with new aggregate. The higher the RAP content, the lower the mixing and compaction temperature of hot recycled asphalt mixture.

3.3. *Effect of RAP Content on Road Performance*

3.3.1. High-Temperature Stability

The calculated results of the rutting test are shown in Table 5. It shows that the dynamic stability (DS) of the hot recycled asphalt mixture met the technical requirements of 800 times/mm according to the Chinese standard JTG F40-2004 [39]. With the increase of RAP content, the dynamic stability of the hot recycled asphalt mixture in the rutting test continues to increase. When the RAP content increases from 30% to 60%, the dynamic stability increased by 38.9%. This was probably because the aged asphalt was softened under the coordination of the rejuvenator and the virgin asphalt, but its viscosity value was still higher than the virgin asphalt, thus improving the rutting resistance of the recycled pavement.

Table 5. Dynamic stability (DS) with different RAP content.

RAP Content (%)	Mixing Temperature (°C)	Compaction Temperature (°C)	DS (Time mm^{-1})	
			Average Value	Specification Requirement
30	175	152	2872	≥800
40	170	150	3216	
50	165	144	3753	
60	155	139	3989	

3.3.2. Moisture Susceptibility

The freeze-thaw splitting and immersion Marshall results of recycled asphalt mixtures with different RAP content are shown in Table 6. It shows that the freeze-thaw splitting tensile strength ratio (TSR) and residual Marshall stability (MS_0) of recycled asphalt mixtures all met the requirement of 75% and 80%. The moisture-induced damage of the hot recycled asphalt mixture decreased with the RAP content increases. It may be that the adhesion of the aged asphalt was reduced due to the addition of large amounts of aged asphalt raw materials, which led to the decrease of moisture-induced damage; when the RAP content reached 60%, the freeze-thaw splitting strength ratio and residual Marshall stability of recycled asphalt mixture were close to the specification value, which indicated that the excessive RAP content would reduce the asphalt cohesion and cause moisture damage, so the proportion of RAP should be controlled.

Table 6. Freeze-thaw splitting tensile strength ratio (TSR) and residual Marshall stability (MS_0) of recycled asphalt mixtures with different RAP content.

RAP Content (%)	Mixing Temperature (°C)	Compaction Temperature (°C)	TSR (%)		MS_0 (%)	
			Test Result	Specification Requirement	Test Result	Specification Requirement
30	175	152	89.6	≥75	87.6	≥80
40	170	150	85.9		86.6	
50	165	144	79.3		83.1	
60	155	139	76.7		81.3	

3.3.3. Low-Temperature Crack Resistance

The low-temperature bending results of recycled asphalt mixtures with different RAP content are shown in Figure 9. Figure 9 presents that with the increase of RAP content, the breaking strain decreased and the bending stiffness modulus increased, which indicated that the addition of RAP would adversely affect the low-temperature crack resistance. Based on the opposite change trend of breaking strain and bending stiffness modulus, the evaluation index of low temperature crack resistance of reclaimed asphalt mixture was the strain energy density of 10 KJ/m^3.

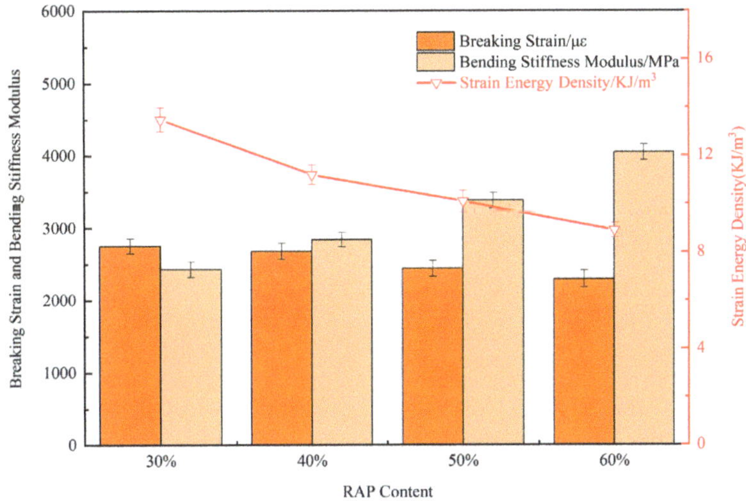

Figure 9. Low-temperature bending results of recycled asphalt mixtures with different RAP content.

3.4. Effect of Rejuvenation Process

The influence of 30% RAP content recycled asphalt mixture with different rejuvenation construction processes on the road performance of asphalt mixture are shown in Figure 10. It can be seen that under the condition of rejuvenation construction process II, the high-temperature stability and moisture-induced damage of recycled asphalt mixture were better than those of rejuvenation construction process I and III. However, the low-temperature crack resistance of recycled asphalt mixture under the condition of rejuvenation construction process II was far lower than that of rejuvenation construction process III.

As can be observed in Figure 10, different rejuvenation processes have different effects on the road performance of the mixture. Therefore, when determining the rejuvenation construction process, it should be selected in combination with the local natural conditions, so that the performance of the recycled asphalt mixture can be adapted to the natural conditions to achieve the purpose of prolonging the service life of the pavement.

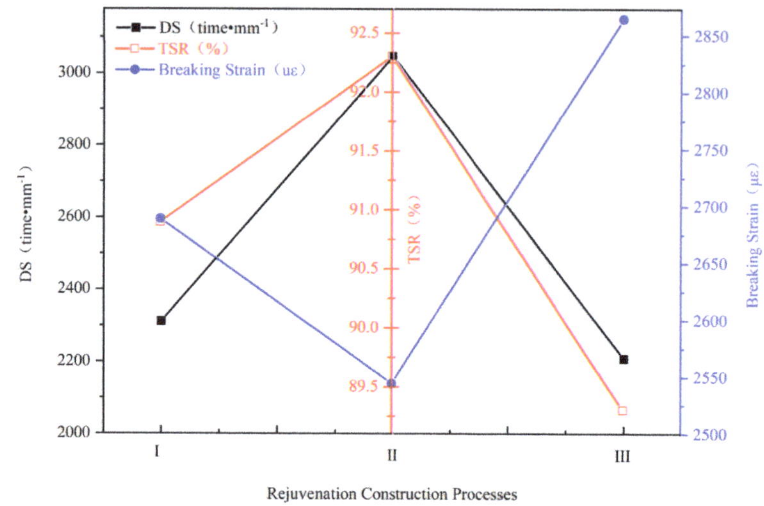

Figure 10. Road performance results of recycled asphalt mixture with different rejuvenation construction processes.

4. Conclusions

The following conclusions can be drawn from this paper:

1. As the content of the rejuvenator increases, the softening point of the recycled asphalt increases, the penetration and ductility decrease and the indexes of R_a, R_q and R_{max} gradually decrease. Based on the penetration, ductility, softening and Nanomorphology analysis of recycled asphalt, the optimum rejuvenator content is 4%.
2. The damage of aggregate can be reduced by milling the pavement under high temperature and ensure the uniformity of the original pavement gradation. For every 10% increase in RAP content, the average mixing and compaction temperature of hot recycled asphalt mixture decreases by 2–6 °C.
3. As the RAP content increases, the high-temperature stability of the hot recycled asphalt mixture increases, the moisture-induced damage and the low-temperature crack resistance decrease. The strain performance density of 10 KJ/m^3 is proposed as the index to evaluate the low-temperature performance of recycled asphalt mixture. From comprehensive road performance results, the optimum percentage of RAP is 30%.
4. High moisture-induced damage and high-temperature stability can be obtained by using RAP + rejuvenator co-heating construction process for recycled asphalt mixture. High low-temperature crack resistance can be obtained by using a RAP + rejuvenator not-heating construction process.

Author Contributions: Data curation, methodology, writing—original draft preparation, T.J.; conceptualization, methodology, writing—review and editing, M.H.; data curation, formal analysis, Q.F.; data curation, investigation, S.M.; supervision, writing—review and editing, X.Y. All authors have read and agreed to the published version of the manuscript.

Funding: This research received no external funding.

Institutional Review Board Statement: Not applicable.

Informed Consent Statement: Not applicable.

Data Availability Statement: All data that support the findings of this study are included within the article.

Conflicts of Interest: The authors declare no competing financial interest.

References

1. Nosetti, A.; P'erez-Madrigal, D.; P'erez-Jim'enez, F.; Martínez, A.H. Effect of the recycling process and binder type on bituminous mixtures with 100% reclaimed asphalt pavement. *Construct. Build. Mater.* **2018**, *167*, 440–448. [CrossRef]
2. Niu, W.G. Situation and development of hot in-place recycling technology of asphalt pavement. *J. China Foreign Highw.* **2019**, *39*, 50–59.
3. Zaumanis, M.; Mallick, R.B.; Frank, R. 100% recycled hot mix asphalt: A review and analysis. *Resour. Conserv. Recycl.* **2014**, *92*, 230–245. [CrossRef]
4. Bressi, S.; Santos, J.; Oreškovi ́c, M.; Losa, M. A comparative environmental impact analysis of asphalt mixtures containing crumb rubber and reclaimed asphalt pavement using life cycle assessment. *Int. J. Pavement Eng.* **2021**, *22*, 524–538. [CrossRef]
5. Milad, A.; Taib, A.M.; Ahmeda, A.G.F.; Solla, M.; Yusoff, N.I.M. A Review of the use of reclaimed asphalt pavement for road paving applications. *J. Teknol.* **2020**, *82*, 35–44. [CrossRef]
6. Jahanbakhsh, H.; Karimi, M.M.; Naseri, H.; Nejad, F.M. Sustainable asphalt concrete containing high reclaimed asphalt pavements and recycling agents: Performance assessment, cost analysis, and environmental impact. *J. Clean. Prod.* **2020**, *244*, 118837. [CrossRef]
7. Zaumanis, M.; Mallick, R.B.; Frank, R. 100% hot mix asphalt recycling: Challenges and benefits. *Transp. Res. Procedia* **2016**, *14*, 3493–3502. [CrossRef]
8. Dughaishi, H.A.; Lawati, J.A.; Bilema, M.; Babalghaith, A.M.; Mashaan, N.S.; Yusoff, N.I.M.; Milad, A. Encouraging sustainable use of RAP materials for pavement construction in Oman: A Review. *Recycling* **2022**, *7*, 35. [CrossRef]
9. Chou, C.P.; Lee, N. A sensitivity study of RAP cost and performance on its life cycle benefits. *Trans. Tech. Publ.* **2013**, *723*, 567–574. [CrossRef]
10. Ma, T.; Huang, X.; Zhao, Y.; Zhang, Y.; Wang, H. Influences of preheating temperature of RAP on properties of hot-mix recycled asphalt mixture. *J. Test. Eval.* **2016**, *44*, 762–769. [CrossRef]
11. Cao, R.J.; Leng, Z.; Hsu, S.C. Comparative eco-efficiency analysis on asphalt pavement rehabilitation alternatives: Hot in-place recycling and milling-and-filling. *J. Clean. Prod.* **2019**, *210*, 1385–1395. [CrossRef]
12. Ma, Y.T.; Polaczyk, P.; Zhang, M.M.; Xiao, R.; Jiang, X.; Huang, B.S. Comparative study of pavement rehabilitation using hot in-place recycling and hot-mix asphalt: Performance evaluation, pavement life prediction, and life cycle cost analysis. *Transp. Res. Record.* **2022**, *2677*, 03611981221099907. [CrossRef]
13. Pan, Y.Y.; Han, D.D.; Yang, T.; Tang, D.; Huang, Y.X.; Tang, N.X.; Zhao, Y.L. Field observations and laboratory evaluations of asphalt pavement maintenance using hot in-place recycling. *Constr. Build. Mater.* **2021**, *271*, 121864. [CrossRef]
14. Ding, L.T.; Wang, X.C.; Zhang, M.Y.; Chen, Z.; Meng, J.Q.; Shao, X.S. Morphology and properties changes of virgin and aged asphalt after fusion. *Constr. Build. Mater.* **2021**, *44*, 207–213. [CrossRef]
15. Yu, J.; Guo, Y.; Peng, L.; Guo, F.; Yu, H. Rejuvenating effect of soft bitumen, liquid surfactant, and bio-rejuvenator on artificial aged asphalt. *Constr. Build. Mater.* **2020**, *254*, 119336. [CrossRef]
16. Ma, Y.T.; Hu, W.; Polaczyk, P.A.; Han, B.Y.; Xiao, R.; Zhang, M.M.; Huang, B.S. Rheological and aging characteristics of the recycled asphalt binders with different rejuvenator incorporation methods. *J. Clean. Prod.* **2020**, *262*, 121249. [CrossRef]
17. Yaseen, G.; Jamal, A.; Almoshageh, M.; Alharbi, F.; Awan, H.H. Performance evaluation of aged asphalt pavement binder through rejuvenators. *Sustainability* **2022**, *14*, 14557. [CrossRef]
18. Chen, A.Q.; Liu, G.Q.; Zhao, Y.L.; Li, J.; Pan, Y.Y.; Zhou, J. Research on the aging and rejuvenation mechanisms of asphalt using atomic force microscopy. *Constr. Build. Mater.* **2018**, *167*, 177–184. [CrossRef]
19. Bilema, M.; Bin Aman, Y.; Hassan, N.A.; Al-Saffar, Z.; Ahmad, K.; Rogo, K. Performance of aged asphalt binder treated with various types of rejuvenators. *Civ. Eng. J.* **2021**, *7*, 502–517. [CrossRef]
20. Zhou, T.; Cao, L.P.; Fini, E.H.; Li, L.W.; Liu, Z.Y.; Dong, Z.J. Behaviors of asphalt under certain aging levels and effects of rejuvenation. *Constr. Build. Mater.* **2020**, *249*, 118748. [CrossRef]
21. Zhao, M.Y.; Shen, F.; Ding, Q.J. Micromechanism of the dispersion behavior of polymer-modified rejuvenators in aged asphalt material. *Appl. Sci.* **2018**, *8*, 1591. [CrossRef]
22. Foroutan Mirhosseini, A.; Kavussi, A.; Tahami, S.A.; Dessouky, S. Characterizing temperature performance of bio-modified binders containing RAP binder. *J. Mater. Civ. Eng.* **2018**, *30*, 04018176. [CrossRef]
23. Ai, X.M.; Cao, J.W.; Feng, D.C.; Gao, L.B.; Hu, W.; Yi, J.Y. Performance evaluation of recycled asphalt mixtures with various percentages of RAP from the rotary decomposition process. *Constr. Build. Mater.* **2022**, *321*, 126406. [CrossRef]
24. Yang, C.; Zhang, J.W.; Yang, F.; Cheng, M.; Wang, Y.F.; Amirkhanian, S.; Wu, S.P.; Wei, M.H.; Xie, J. Multi-scale performance evaluation and correlation analysis of blended asphalt and recycled asphalt mixtures incorporating high RAP content. *J. Clean. Prod.* **2021**, *317*, 128278. [CrossRef]
25. Khan, M.Z.H.; Koting, S.; Katman, H.Y.B.; Ibrahim, M.R.; Babalghaith, A.M.; Asqool, O. Performance of high content reclaimed asphalt pavement (RAP) in asphaltic mix with crumb rubber modifier and waste engine oil as rejuvenator. *Appl. Sci.* **2021**, *11*, 5226. [CrossRef]
26. Shannon, C.; Mokhtari, A.; Lee, H.D.; Tang, S.; Williams, C.; Schram, S. Effects of high reclaimed asphalt pavement content on the binder grade, fatigue performance, and mix design. *J. Mater. Civ. Eng.* **2017**, *29*, 04016218. [CrossRef]

27. Yu, B.; Gu, X.Y.; Wu, M.; Ni, F.J. Application of a high percentage of reclaimed asphalt pavement in an asphalt mixture: Blending process and performance investigation. *Road Mater. Pavement Des.* **2017**, *18*, 753–765. [CrossRef]
28. Bilema, M.; Aman, M.Y.; Hassan, N.A.; Memon, Z.A.; Omar, H.A.; Yusoff, N.I.M.; Milad, A. Mechanical performance of reclaimed asphalt pavement modified with waste frying oil and crumb rubber. *Materials* **2021**, *14*, 2781. [CrossRef]
29. Yin, P.; Pan, B.F. Effect of RAP content on fatigue performance of hot-mixed recycled asphalt mixture. *Constr. Build. Mater.* **2022**, *328*, 127077. [CrossRef]
30. Zhu, J.Q.; Ma, T.; Fan, J.W.; Fang, Z.Y.; Chen, T.; Zhou, Y. Experimental study of high modulus asphalt mixture containing reclaimed asphalt pavement. *J. Clean. Prod.* **2020**, *263*, 121447. [CrossRef]
31. Zaumanis, M.; Arraigada, M.; Poulikakos, L.D. 100% recycled high-modulus asphalt concrete mixture design and validation using vehicle simulator. *Construct. Build. Mater.* **2020**, *260*, 119891. [CrossRef]
32. Ma, Y.T.; Polaczyk, P.; Park, H.; Jiang, X.; Hu, W.; Huang, B.S. Performance evaluation of temperature effect on hot in-place recycling asphalt mixtures. *J. Clean. Prod.* **2020**, *277*, 124093. [CrossRef]
33. Li, X.L.; Lv, X.C.; Zhou, Y.H.; You, Z.P.; Chen, Y.L.; Cui, Z.J.; Diab, A. Homogeneity evaluation of hot in-place recycling asphalt mixture using digital image processing technique. *J. Clean. Prod.* **2020**, *258*, 120524. [CrossRef]
34. Xie, Z.X.; Rizvi, H.; Purdy, C.; Ali, A.; Mehta, Y. Effect of rejuvenator types and mixing procedures on volumetric properties of asphalt mixtures with 50% RAP. *Construct. Build. Mater* **2019**, *218*, 457–464. [CrossRef]
35. Lei, Y.; Wang, H.N.; You, Z.P.; Wang, Y.X.; Jiang, X.; Gao, J.F. Effect of mixing sequence on compaction property of hot recycled asphalt mixtures. *Eng. J. Wuhan Univ.* **2019**, *52*, 317–323.
36. Lei, Y.; Wang, H.N.; Ji, C.J.; You, Z.P. Effect of mixing sequence on high temperature performance of hot recycled asphalt mixture. *J. Jiangsu Univ. (Nat. Sci. Ed.)* **2017**, *38*, 367–372.
37. Chen, Q.; Wang, C.H.; Li, Y.W.; Feng, L.; Huang, S. Performance development of polyurethane elastomer composites in different construction and curing environments. *Construct. Build. Mater* **2023**, *365*, 130047. [CrossRef]
38. Transport, M.O. *Standard Test Methods of Bitumen and Bituminous Mixtures for Highway Engineering*; China Communications Press: Beijing, China, 2011.
39. Transport, M.O. *Technical Specifications for Construction of Highway Asphalt Pavements*; China Communications Press: Beijing, China, 2004.

sustainability

Article

Research on a New Loading Method for Nano TiO_2 Photocatalytic Asphalt Pavement

Jingxiao Shu [1,2], Xiaoyang Wang [3,*], Bo Yang [3] and Xiaofeng Wang [3]

1 School of Hydraulic and Civil Engineering, Zhengzhou University, Zhengzhou 450001, China
2 Henan Expressway Network Management Center, Zhengzhou 450046, China
3 Henan Communications Planning & Design Institute Co., Ltd., Zhengzhou 451460, China
* Correspondence: xywanghn@163.com

Abstract: The main goal of our work was to study a new loading method for photocatalytic asphalt pavement that could effectively solve the problems of photocatalytic degradation efficiency and durability. We adhered nano TiO_2 particles to the microscopically textured structure on the surface of glass microbeads by cold alkaline corrosion and high-temperature adhesion technology. We observed good adhesion of nano TiO_2 on glass microbeads by a microscopic performance characterization of the composites. The improvement in the light transmittance of the composite material improved the catalytic efficiency of nano TiO_2 to a certain extent. Three different groups were established to verify the durability of the nano TiO_2 loading method. The result shows that the exhaust gas degradation rate of the spray embedding group did not decrease significantly with the increase in road friction time. Our research provides a new idea for the design of exhaust degradation pavement.

Keywords: environmental protection; new loading method; microscopic characterization; photocatalytic asphalt pavement

Citation: Shu, J.; Wang, X.; Yang, B.; Wang, X. Research on a New Loading Method for Nano TiO_2 Photocatalytic Asphalt Pavement. *Sustainability* **2022**, *14*, 11977. https://doi.org/10.3390/su141911977

Academic Editors: Chaohui Wang, Dawei Wang, Kai Liu and Qian Chen

Received: 10 August 2022
Accepted: 8 September 2022
Published: 22 September 2022

1. Introduction

It is generally acknowledged that vehicle exhaust emissions, which contain many harmful gases, such as NOx, that may form acid rain, are a major problem in environmental governance around the world. Various measures have been taken to alleviate the negative effects of exhaust gas, but most of them have achieved little. During the past decade, there has been an increasing interest in photocatalytic exhaust gas degradation materials. Nano TiO_2, as a classic photocatalyst, is an environmentally friendly material that can effectively purify automobile exhaust.

Japan is one of the first countries to promote photocatalytic materials in the field of road engineering. In an earlier study, some new loading technologies emerged that not only allowed photocatalytic materials to be recycled but also had more efficient air purification capabilities. Hisanaga T [1] studied the catalytic degradation effect of pure nano TiO_2 and zeolite-nano TiO_2 composites on benzene and found that increasing the humidity of gas-phase benzene samples was beneficial for improving the catalytic activity of nano TiO_2. Bing yu Jia [2] loaded nano TiO_2 onto glass fibers, which confirmed the possibility of recycling photocatalysts; the photocatalytic process of the composite photocatalyst exhibited a higher degradation rate than pure nano TiO_2. Japan's Mitsubishi Corporation [3] has developed a new type of coating, with nano TiO_2 as the main component, that can absorb the nitrogen oxides found in automobile exhaust with a rate of high removal, high efficiency, and good durability.

Hassan [4] made a keynote speech at the annual meeting of the Association of Asphalt Paving Technologists (AAPT), pointing out that nano TiO_2 can be used as a pavement coating to break down car exhaust, which would revolutionize the environmental performance of traditional hot mix asphalt. Before that, he built the nation's first test road for air-purifying photocatalytic asphalt pavement on the campus of Louisiana State University [5,6]. It is important to highlight that air-purifying photocatalytic asphalt pavement can effectively eliminate 31~55% of nitrogen oxides and 4~20% of sulfur dioxide pollutants in the air, which has an obvious effect on the purification of car exhaust. However, one remaining challenge is that the durability of nano TiO_2 photocatalytic nanoparticle coatings still needs to be studied further, according to the aforementioned report.

Although nano TiO_2 photocatalytic oxidation technology provides a good way to degrade air pollutants, it continues to be challenging to develop highly active methods of application of nano TiO_2 for asphalt pavement exhaust degradation. It is worthwhile to mention that the nano TiO_2 loading methods used in road engineering mainly consist of mixing nano TiO_2 powder into pavement materials or spraying nano TiO_2 aqueous slurry onto the pavement [7]. In the first case, since the doped nano TiO_2 is wrapped by the pavement material, which affects its contact with sunlight and polluting gases, the nano TiO_2 is wasted due to its inability to exert photocatalytic activity [8]. In the second case, the nano TiO_2 aqueous slurry sprayed onto the surface has difficulty bonding firmly with the pavement material, has a short service life, and affects the anti-skid performance of the pavement [9,10].

In this paper, we propose a new nano-loading method for photocatalytic asphalt pavement that can effectively solve the problems of photocatalytic degradation efficiency and durability.

In summary, this paper makes the following contributions:

(1) We adhere nano TiO_2 particles to the microscopically textured structure on the surface of glass microbeads (the main components are Na_2O, B_2O_3, and SiO_2) by cold alkaline corrosion and high-temperature adhesion technology.
(2) We perform a microscopic performance characterization of the composites after loading.
(3) We design an asphalt pavement structure for carrying photocatalytic composites.
(4) We evaluate the durability and exhaust gas degradation efficiency of the new photocatalytic asphalt pavement.

The rest of this paper is organized as follows: Section 2 presents the specific implementation steps for loading nano TiO_2 particles onto glass microbeads, as well as the micro-properties of the composites. In Section 3, we present a semi-flexible pavement design method, based on a macro-porous parent asphalt mixture, that acts as the carrier of nano TiO_2 composites by controlling the surface pore structure reserved after grouting. Section 4 provides detailed experiments to test the durability of the semi-flexible pavement and presents the data analysis for the degradability of vehicle exhaust after the abrasion test. Finally, Section 5 concludes the paper.

2. Materials, Procedures, and Methodologies

2.1. Attempt to Load Nano TiO_2 onto the Surface of Glass Microbeads

Breathing fog, before Widawski (1994) cast polymer solutions onto substrates and discovered a method for the formation of an ordered porous film, was originally a very common and annoying natural phenomenon in medicine and metallurgy. Since this discovery, it has evolved into a remarkable self-assembly strategy for fabricating porous structures with pore sizes ranging from several nanometers to hundreds of nanometers. Inspired by this principle, we propose a new method for loading nano TiO_2 by manufacturing numerous pore-like structures in glass microbeads to enhance the light transmission properties of the material.

Nano TiO_2 and glass microbeads were selected for use in this article. The basic properties of nano TiO_2 and glass microbeads are shown in Tables 1 and 2, respectively.

Here, the nano TiO_2 particles are preloaded onto the surface of glass microbeads treated by cold-alkaline corrosion (CAC). Then, nano TiO_2 particles are adhered to the microscopically textured structure on the surface of the glass microbeads by high-temperature adhesion (HTA) to form a stable and firm adhesion structure by using the physical properties of B_2O_3, which melts at 450 °C. The specific implementation processes are as follows:

(a) a saturated $Ca(OH)_2$ solution is prepared in a constant-temperature water tank (20 °C), and the excess $Ca(OH)_2$ solid particles are filtered out; (b) the glass microbeads are soaked in the saturated $Ca(OH)_2$ solution for 12 h; (c) we filter (sieve aperture 0.075 mm) and wash the soaked glass microbeads and dry them in an oven at 105 °C; (d) we use a high-speed mixer (600 r/min, 15 min) to stir until we attain a nano TiO_2 solution with a concentration of 0.5%; (e) we put glass microbeads into the nano TiO_2 solution, stirring for 30 min at the same speed, and then we repeat step (c); (f) we calcine the dried glass microbeads in a muffle furnace at 450 °C for 2 h; finally, after cooling, the loading process of the nano TiO_2 composite is finished.

Table 1. Technical properties of nano TiO_2.

Technical Indexes	Unit	Test Results
Exterior	/	white powder
Particle size	nm	10
TiO_2 content	%	>99.5
Melting point	°C	1830–1850
Specific surface area	m^2/g	80
Density	g/cm^3	0.3
Surface properties	/	hydrophilic

Table 2. Technical properties of glass microbeads.

Technical Indexes	Unit	Test Results
Water soluble		not soluble in water
Color	/	pure white
pH	/	9.5
Compressive strength	MPa	1.7–124
Dielectric constant	/	1.2–2
Density	g/cm^3	0.18–0.21
Particle size range	μm	15–150

2.2. *Surface Composition and Morphology*

Considering that nano TiO_2 and glass microbeads may have poor adhesion during high-speed stirring, this article focuses on a surface composition and morphology analysis to observe whether the two achieve a good adhesion effect.

Whether there is a successful reaction of nano TiO_2 with glass microbeads can be established by analyzing the surface composition before and after the treatment with nano TiO_2, glass microbeads, and the composite material. The scanning electron microscopy (SEM) and X-ray diffraction (XRD) test results are presented in Figures 1 and 2.

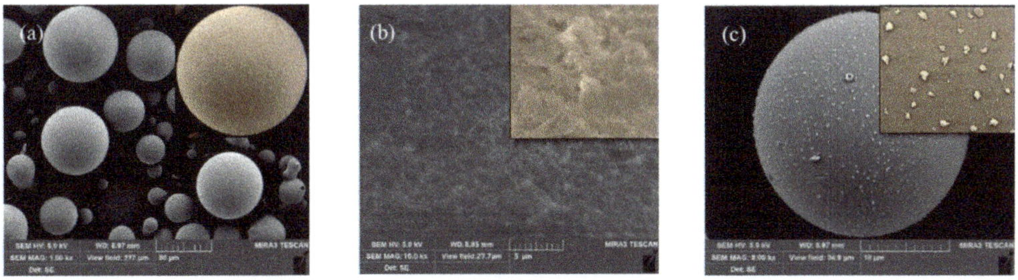

Figure 1. (**a**) The glass microbeads(50 μm); (**b**) cold-alkaline corrosion(5 μm); (**c**) the nano TiO_2 composite(50 μm).

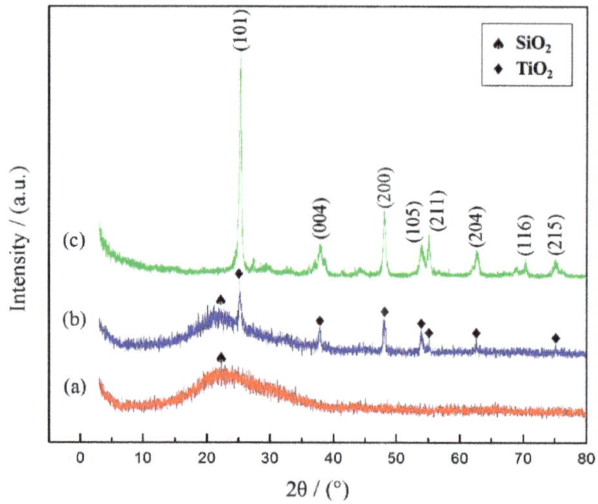

Figure 2. (**a**) The glass microbeads; (**b**) the composite material; (**c**) nano TiO_2.

Figure 1a shows the surface of the glass microbeads is very smooth before being soaked in a saturated $Ca(OH)_2$ solution. After the soaking process, an irregular gully-like texture (the width is 20 nm–50 nm, and the depth is about 10 nm) appears on the surface of the glass microbeads, which provides a good channel for loading nano TiO_2 (the average particle size is 10 nm), as shown in Figure 1b. It is worthwhile to mention that the main component of glass microbeads is sodium borosilicate, which is an amorphous material composed of three components: SiO_2, Na_2O, and B_2O_3. The melting points of the three main components are 1650 °C, 1132 °C, and 445 °C, respectively, which means the surface of glass microbeads produces molten B_2O_3 at a temperature of 445 °C. According to this theory, we calcine the soaked glass microbeads in a muffle furnace at 450 °C for 2 h in order to adhere nano TiO_2 onto the trench-like texture created by the molten B_2O_3. Figure 1c shows the adhesion-treated photocatalytic composites with a large amount of nano TiO_2 attached to the surface.

Figure 2 shows the crystal energy spectrum results for nano TiO_2, glass microbeads, and the composite material. The X-axis describes crystals at the test point according to the energy intensity of the characteristic spectral lines, whereas the Y-axis reflects the relative count of the crystals at the test point. Figure 2 shows the results for glass microbeads (a), the composite material (b), and nano TiO_2 (c), according to the diffraction parameter of the Joint Committee on Powder Diffraction Standards (JCPDS), Card No. 21-1272. Seven simple anatase nano TiO_2 characteristic peaks, namely (101), (004), (200), (105), (211), (204), and (215), can be found in (b,c), which means that the nano TiO_2 has been embedded on the surface of the glass microbeads. It must also be mentioned that the characteristic peak (116) was not found in (b), which means that the small intensity of this peak may be masked by the amorphous diffraction peak of the glass microbeads.

2.3. Light Transmittance Analysis

Glass microbeads are a chemically stable mixture and are usually used as a reinforcing agent for plastics. In addition, they have the properties of reflection and diffusion; as such, they are often used in road signs and pavement markings in the field of traffic to enhance light sources and warnings. This article focuses on the photocatalytic degradation rate of nano TiO_2, for which sufficient light is a catalytic condition. This section focuses on a test of the light transmittance of nano TiO_2 composites by UV-Vis Spectrophotometer. Figure 3 shows that the average transmittance of the glass microbead–nano TiO_2 composite material is 94.5%, which is 26.7% higher than that of pure nano TiO_2. The result shows that the composite material has stronger light transmittance than pure nano TiO_2. It must also be mentioned that the blue line and the red line have a similar inflection point that appears around the wavelength of 380 nm, which offers additional proof of the successful production of the glass microbeads–nano TiO_2 composites.

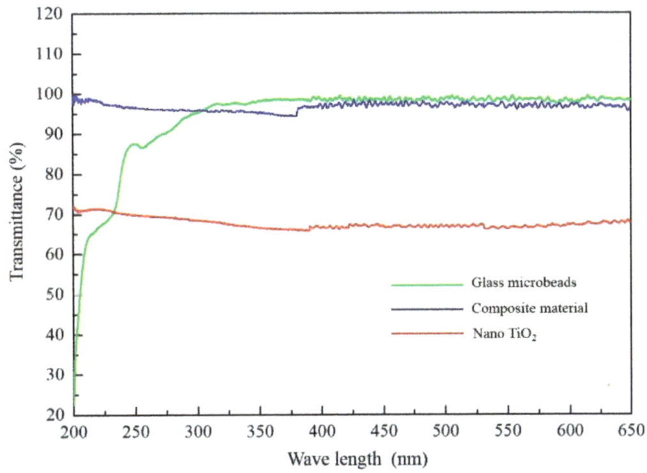

Figure 3. Light transmittance analysis.

3. Durable Pavement Structure Design

This section focuses on durable pavement structure design to reduce the loss of exhaust gas degradation materials due to friction with tires on the road. This paper is inspired by a type of macro-porous asphalt mixture that can provide sufficient construction depth to carry nano TiO_2 composites [11–13]. This structure is divided into two parts; one part is the macro-porous parent asphalt mixture, and the other part is the cement-filling slurry material. As such, it offers important advantages over dense-graded asphalt mixtures; it possesses sufficient construction depth to reduce the friction between the tire and the photocatalytic material when they are in direct contact, while improving the durability of the nano TiO_2 composites. It is worthwhile to mention that alkaline cement slurry can neutralize the acid product (nitric acid) of the photocatalytic degradation of gas pollutants, forming a sustainable road exhaust degradation cycle under the scouring of rainwater, which is another significant advantage.

The asphalt pavement structure design steps are as follows: (1) select SFAC-13 with large pores as the master mix gradation, and select the optimal design void ratio of 24%; (2) configure the proportion of cement grouting materials and then determine the optimal reserved grouting depth of the master mix according to the test results.

3.1. Gradation Design of Matrix Asphalt Mixtures

The gradation used in the design of the asphalt mixture in this paper refers to the relevant requirements in the "Technical Specification for Design and Construction of Perfusion Semi-flexible Pavement" (DB11/T 1817-2021). The gradation composition and gradation curve of the base asphalt mixture are shown in Table 3 and Figure 4, respectively.

According to the gradation of SFAC-13 given in Table 1, the Marshall test procedure is used to determine the optimum asphalt content for the asphalt mixture, according to the "Highway Asphalt Pavement Construction Technical Specifications" (JTJF40-2004) requirements. The optimum content of the asphalt binder is 3.4%, and the target void ratio is 24%.

Table 3. The aggregate gradation of SFAC-13.

Sieve Size (mm)	16	13.2	4.75	2.36	0.6	0.3	0.15	0.075
Upper limit of gradation (%)	100	100	30	22	15	12	8	6
Lower limit of gradation (%)	100	90	10	5	4	3	3	1
Synthetic gradation (%)	100	95	20	13.5	9.5	7.5	5.5	3.5

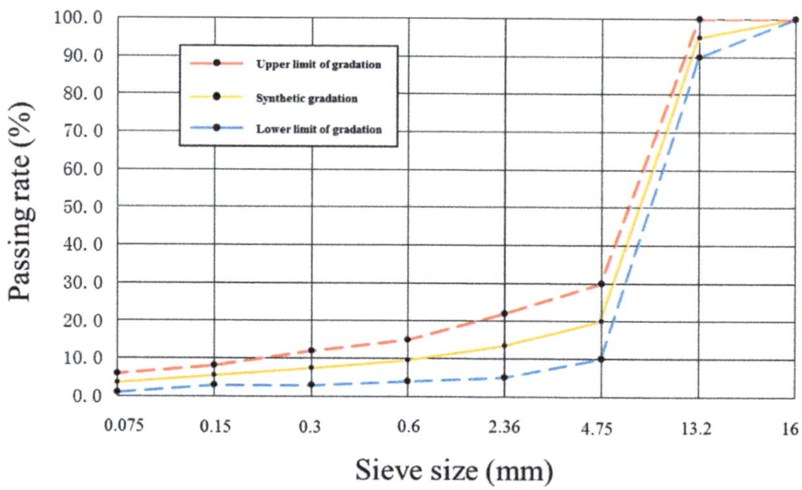

Figure 4. The aggregate gradation of SFAC-13.

3.2. Optimum Depth Selection of Cement Mortar Grouting

The cement mortar selected in this paper is 42.5 R ordinary Portland cement. Performance test indicators include the initial setting time, final setting time, compressive strength, and flexural strength, according to the "Testing Methods of Cement and Concrete for Highway Engineering" (JTG3420-2020). The test results are shown in Table 4.

Table 4. The test results for 42.5 R ordinary Portland cement mortar.

Detection Indicator	Unit	Test Results	Requirement
Initial setting time	min	183	≥90
Final setting time	min	241	≤600
Compressive strength (7D)	MPa	38.45	10–30
Flexural strength (7D)	MPa	7.80	>2

In contrast to general flexible pavements, semi-flexible pavement performance is characterized by a mix of flexibility and rigidity; it has not only the advantages of asphalt pavement in terms of noise reduction, sound absorption, driving comfort, etc., but also the advantages of strong bearing capacity and anti-rutting that are typical of cement pavement. Yet, the biggest problem with semi-flexible pavement is the coordinated deformation ability of its dissimilar materials, which may cause cracks in the road [14,15]. Thus, the strength stability of the coordinated deformation of dissimilar materials is paid more attention when selecting the grouting depth of the cement grout.

Considering that the maximum nominal particle size of the aggregate is 13.2 mm, we take the largest nominal particle size of the aggregate as the maximum reserved grouting depth and decrease it with a gradient of 20% to find the best reserved grouting depth. Before grouting the asphalt mixture, we fix the specimens on the cement vibrating table, which is sealed on the bottom and sides with plastic wrap and foil. We grout the specimens while vibrating for 90 s so that the cement mortar can be fully poured into the connected gaps of the specimen. After the perfusion completes, the specimens are left to stand for about 30 min. Then, we use a rubber rake to wipe off the excess cement mortar on the surface of the specimens and place the specimens in a curing box at 20 °C for 6 days.

In order to obtain the best reserved depth of the specimens, we focus on a series of mechanical property tests for semi-flexible specimens with different reserved grouting depths, including Marshall stability, high-temperature stability, low-temperature crack resistance, and water stability. The test results are shown in Figure 5.

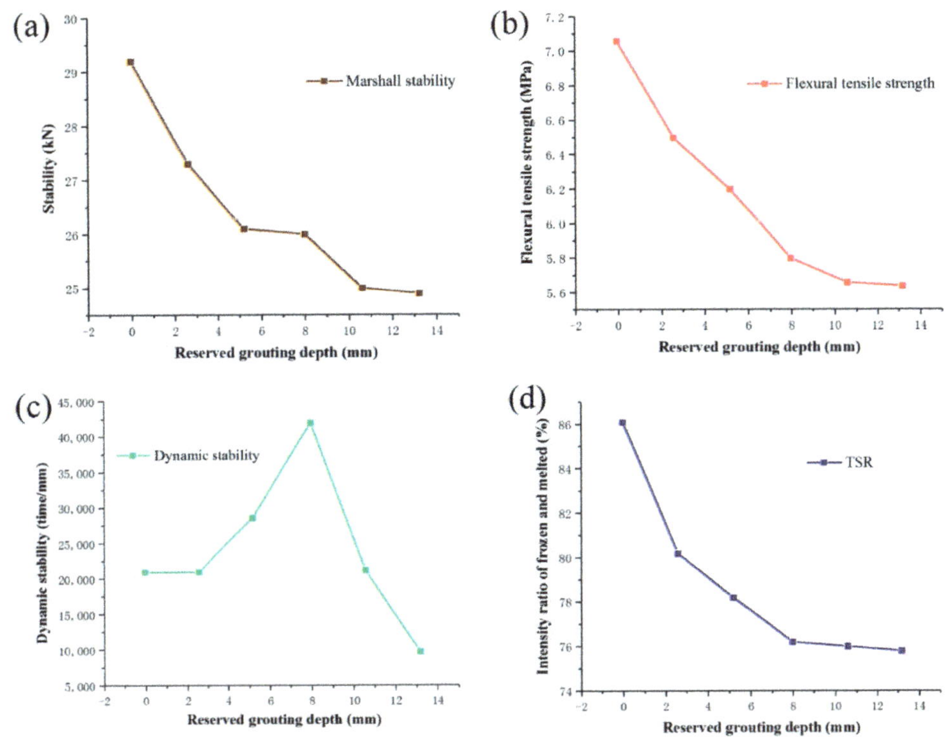

Figure 5. Relationship between properties test results and the grouting depth: (**a**) Marshall stability; (**b**) the flexural tensile strength; (**c**) dynamic stability; (**d**) TSR.

As can be seen from (a,b) of Figure 5, the attenuation law of Marshall stability and the flexural tensile strength show that the curves no longer have obvious attenuation when they reach 10.56 mm, at which point they tend to be stable. From the law of dynamic stability (c), it can be concluded that the semi-flexible asphalt mixture reaches its peak value when the reserved grouting depth is 7.92 mm. In the freeze–thaw splitting test results (d), the reserved grouting depth tends to be stable between 8 mm and 10 mm. Considering the test results alone, the optimal grouting depth of cement grout should be controlled between 7.92 mm and 10.56 mm. Considering that the glass microbead–nano TiO_2 composite materials need to reserve a certain paving space when pouring cement mortar, improvements in appropriately adjusting the optimal grouting depth of cement grout should be considered.

4. Evaluation of the Degradation Efficiency of Automobile Exhaust Gas

Given the observed problem with the durability of nano TiO_2 photocatalytic nanoparticle coatings, this section focuses on the wear resistance of composite particles in pavement, which is assessed by the change in the degradation efficiency of automobile exhaust gas before and after the abrasion test.

In this experiment, we set up a stirring group (SG), in which we replace mineral powder with the same quantity of composite materials during the mixing process, a spray embedding group (SEG), in which we spray the composite materials on the surface of the unset grout, and an aqueous coating group (ACG), in which we configure the composite water solution and spray applied to the surface of the fully set grout for different durations of abrasion to verify the durability of the new nano TiO_2 loading method. In order to facilitate the analysis of the test results, the parameters of the nano TiO_2 composites in the three groups are uniformly set to 40 g (the mineral powder in the stirring group is 40 g). We conduct the wear test in a dry environment, rubbing each group of specimens under standard wheel pressure (0.7 MPa). After a phased test (2 h), we use a blower to clean up the dust on the surface of the specimens and then conduct vehicle exhaust gas degradation tests to complete the phased testing. The entire test trial is divided into six stages, for a total of 12 h.

Under sunlight, highly catalytically active groups can be generated on the surface of nano TiO_2, which has a strong redox ability. Yet, the photocatalytic reaction process of photocatalysts requires enough excited photons functioning as catalysts to provide sufficient energy [16–18]. Generally, the ultraviolet wavelength in sunlight can meet the excitation requirements of photocatalysts. Therefore, in the simulated degradation reaction experiments, UV irradiation intensity is a key factor in the reaction rate and degradation ability. However, under outdoor conditions, UV light intensity changes due to the weather conditions, especially humidity and cloud cover, which may lead to certain errors in the test results. Taking the above observations into account, we carry out the test indoors under 600 W/m^2 UV light irradiation.

The vehicle exhaust gas degradation test is carried out in a dark room. The specimens are placed in a sealed glass box that has three UV-emitting lamps on the top to adjust the intensity of UV rays. In order to completely avoid the influence of sunlight on the test results, we cover the gas reaction box with a completely effective shading cloth during the test. The sample preparation and the test steps are shown in Figure 6.

The residual degradation ability of the specimens mentioned in this paper refers to the degradation ability of the specimen with respect to automobile exhaust after wheel wear. After preliminary testing, the residual degradation ability of the three groups of specimens is tested under the irradiation of UV light at 600 W/m^2 for 30 min because other light intensities make the test either too long or too slow. The average degradation ability results for nitrogen oxides, carbon oxides, and sulfur oxides in automobile exhaust are shown in Table 5.

Table 5. Evaluation of residual degradation ability of specimens under UV light at 600 W/m^2.

Test Time (h)	Stage	Spray Embedding Group (%)			Aqueous Coating Group (%)			Stirring Group (%)		
		NO_X	CO_X	SO_X	NO_X	CO_X	SO_X	NO_X	CO_X	SO_X
0	/	85.7	9.1	18.9	86.6	6.9	21.9	2.48	0	0.72
2	1	86.4	8.2	16.8	71.4	3.2	12.4	2.56	0	0.66
4	2	86.3	7.9	16.2	69.5	3.1	12.6	2.13	0	0.64
6	3	84.8	7.6	16.1	67.9	2.9	12.8	2.24	0	0.68
8	4	85.6	7.8	15.8	63.3	3.1	11.9	2.11	0	0.52
10	5	84.9	7.7	15.5	62.6	3.0	11.3	2.31	0	0.60
12	6	84.6	7.7	15.6	60.3	2.9	11.5	1.95	0	0.58

After the test specimens for different paving methods are tested, there are significant differences in the degree of degradation capacity. Figure 7 shows that the stirring group has a poor degradation efficiency of only 1.09%. This is because the composite material is mixed into the interior of the specimen, replacing the mineral powder, and, thus, cannot be exposed to the external light source to participate in the photocatalytic reaction. After the first stage of testing for the aqueous coating group, the degradation efficiency with respect to harmful gases, such as nitrogen oxides, carbon oxides, and sulfur oxides, decreases by 17.49%, 53.62%, and 43.25%, respectively. After the fourth stage, the degradation rate is gradually stabilized. The reason it still partially retains the ability to degrade exhaust is that some composite material is embedded in the surface texture of the asphalt mixture. In contrast, in the spray embedding group, the exhaust gas degradation rate is almost not attenuated for its good durability.

Figure 6. Gas degradation test steps: (**a**) specimen preparation; (**b**) abrasion test on the specimens; (**c**) exhaust gas analyzer debugging; (**d**) gas degradation test.

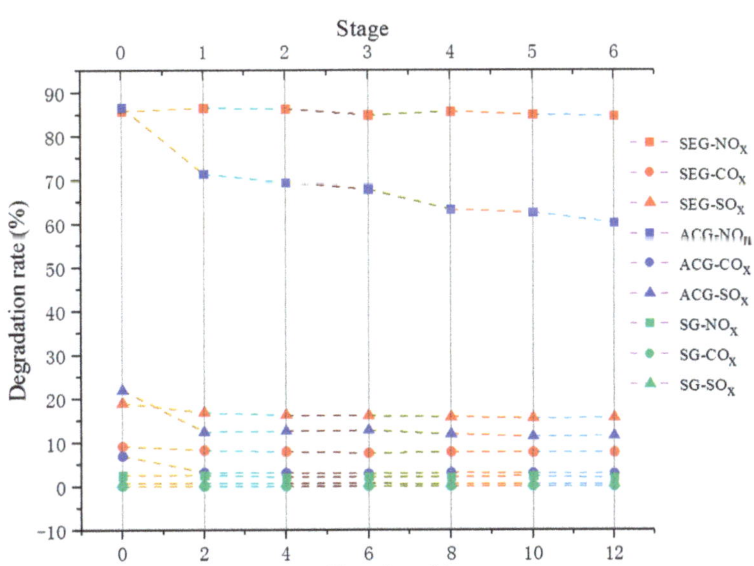

Figure 7. Test results of the exhaust gas degradation test.

5. Conclusions

A new nano TiO_2 loading method is proposed that can effectively improve the refractive index of illumination and the durability of materials.

We observed good adhesion of nano TiO_2 on glass microbeads via the microscopic performance characterization of composites. The improvement in the light transmittance of the composite material improves the catalytic efficiency of nano TiO_2 to a certain extent.

The structure of the macro-porous asphalt mixture designed in this paper protects the composite material from tire friction loss and improves the durability of the material to a certain extent.

Author Contributions: Formal analysis, X.W. (Xiaoyang Wang); Methodology, J.S.; Resources, B.Y. and X.W. (Xiaofeng Wang). All authors have read and agreed to the published version of the manuscript.

Funding: This work was supported by the following projects or agents: the Transportation Science and Technology Achievements Promotion Project of Henan Provincial Department (2021T2), the Transportation Science and Technology Achievements Promotion Project of Henan Provincial Department (2021T8), and the Transportation Technology Project of Henan Provincial Department (2021G3).

Informed Consent Statement: Informed consent was obtained from all subjects involved in the study. Written informed consent has been obtained from the patients to publish this paper.

Conflicts of Interest: The authors declare no conflict of interest.

References

1. Hisanaga, T.; Tanaka, K. Photocatalytic degradation of benzene on zeolite-incorporated TiO_2 film. *J. Hazard. Mater.* **2002**, *93*, 331–337. [CrossRef]
2. Qiao, X.J.; Li, P.; Wen, L. Study on the effect of nano-titanium dioxide environmental protection coatings on degrading nitrogen oxides in automobile exhaust. *Constr. Technol.* **2014**, *43*, 664–666.
3. Jia, B.Y.; Duan, L.Y.; Ma, C.L.; Wang, C.M. Characterization of TiO_2 loaded on activated carbon fibers and its photocatalytic reactivity. *Chin. J. Chem.* **2007**, *25*, 553–557. [CrossRef]
4. Hassan, M.; Mohammad, L.N.; Dylla, H.; Asadi, S.; Cooper, S. Laboratory and field evaluation of sustainable photocatalytic asphalt pavements. *AAPT* **2012**, *81*, 1–20.
5. Photocatalyst Coating Agents, a Proposal of Advanced Environmental Purification Technologies [EB/OL]. 2004. Available online: http://www.khi.co.jp/folium (accessed on 2 May 2022).
6. Qian, C.X.; Zhao, L.F.; Fu, D.F. Research on photocatalytic activity of road surface materials. In Proceedings of the 17th ASCE Engineering Mechanics Division Conference, University of Delaware, Newark, DE, USA, 13–16 June 2004.
7. Leng, Z.; Yu, H. Novel method of coating titanium dioxide onto asphalt mixture based on the breath figure process for air-purifying purpose. *J. Mater. Civ. Eng.* **2016**, *28*, 1–7. [CrossRef]
8. Zhong, L.; Haghighat, F. Photocatalytic air cleaners and materials technologies-Abilities and limitations. *Build. Environ.* **2015**, *91*, 191–203. [CrossRef]
9. Zhu, T.L. *Preparation of Modified Nano-TiO_2 Composites and Research on Durability and Functional Finishing of Textiles*; Shanghai University of Engineering and Technology: Shanghai, China, 2016.
10. Hassan, M.; Mohammad, L.N.; Asadi, S.; Dylla, H.; Cooper, S. Sustainable photocatalytic asphalt pavements for mitigation of nitrogen oxide and sulfur dioxide vehicle emissions. *J. Mater. Civ. Eng.* **2013**, *25*, 365–371. [CrossRef]
11. Zhang, X.N.; Wang, S.H.; Wu, K.H.; Wang, D.Y. CAVF method for composition design of asphalt mixture. *Highway* **2001**, *12*, 17–21.
12. Lai, F.; Huang, Z.; Guo, F. Noise reduction characteristics of macro porous asphalt pavement based on a weighted sound pressure level sensor. *Materials* **2021**, *14*, 4356. [CrossRef]
13. Chen, Q.; Wang, C.; Yu, S.; Song, Z.; Fu, H.; An, T. Low-temperature mechanical properties of polyurethane-modified waterborne epoxy resin for pavement coating. *Int. J. Pavement Eng.* **2022**, 1–13. [CrossRef]
14. Zhang, J.; Cai, J.; Pei, J.; Li, R.; Chen, X. Formulation and performance comparison of grouting materials for semi-flexible pavement. *Constr. Build. Mater.* **2016**, *115*, 582–592. [CrossRef]
15. Pei, J.; Cai, J.; Zou, D.; Zhang, J.; Li, R.; Chen, X.; Jin, L. Design and performance validation of high-performance cement paste as a grouting material for semi-flexible pavement. *Constr. Build. Mater.* **2016**, *126*, 206–217. [CrossRef]
16. Wang, Y.; He, Y.; Lai, Q.; Fan, M. Review of the progress in preparing nano TiO_2: An important environmental engineering material. *J. Environ. Sci.* **2014**, *26*, 2139–2177. [CrossRef] [PubMed]

17. Kwon, S.; Fan, M.; Cooper, A.T.; Yang, H. Photocatalytic applications of micro-and nano-TiO_2 in environmental engineering. *Crit. Rev. Environ. Sci. Technol.* **2008**, *38*, 197–226. [CrossRef]
18. Macwan, D.P.; Dave, P.N.; Chaturvedi, S. A review on nano-TiO_2 sol–gel type syntheses and its applications. *J. Mater. Sci.* **2011**, *46*, 3669–3686. [CrossRef]

sustainability

MDPI

Article

Laboratory Evaluation of Dynamic Characteristics of a New High-Modulus Asphalt Mixture

Haiwei Zhang [1], Xingwang Yang [2,*], Yan Li [3], Qilong Fu [1] and Huayu Rui [1]

[1] School of Civil Engineering and Architecture, Zhengzhou University of Aeronautics, Zhengzhou 450046, China
[2] Henan GOLDROAD Industrial Group Co., Ltd., Xuchang 461000, China
[3] School of Civil Engineering and Architecture, Nanyang Normal University, Nanyang 473061, China
* Correspondence: zhenghang0789@163.com

Abstract: With the rapid increase in traffic volume and heavy-duty vehicles, rutting has become one of the most serious problems threatening the service quality and life of asphalt pavement. High-modulus asphalt concrete is a promising method to overcome this problem, contributing to the sustainable development of asphalt pavement. In this study, a new composite high-modulus agent (CHMA)-modified asphalt binder and mixture were prepared, and their dynamic mechanical characteristics were investigated by the dynamic shear rheometer, dynamic modulus test, wheel tracking test, frequency sweep test at a constant height (FSCH), and repeated shear test at a constant height (RSCH) to comprehensively evaluate its high-temperature stability. Test results showed that the rheological property of the CHMA-modified asphalt binder was similar to that of low-graded asphalt binder, implying that it had a strong potential in resisting deformation. The dynamic modulus of AC-20(CHMA) was 19,568 MPa at 15 °C and 10 Hz condition, meeting the requirement for the high-modulus asphalt mixture (higher than 14,000 MPa). The dynamic stability of AC-20(CHMA) was 8094 times/mm, lower than that of AC-20(20#), but remarkably higher than that of AC-20(SBS). AC-20(20#) and AC-20(CHMA) both showed strong shear resistance according to the FSCH test results. Under the repeated shear loadings, the growth rate of the shear strain increased rapidly in the primary stage, and then slowed down gradually, finally reaching a constant growth rate. The shear slope of AC-20(CHMA) was between that of AC-20(20#) and AC-20(SBS), demonstrating that its resistance to repeated shear loadings was superior to AC-20(SBS), although slightly weaker than AC-20(20#). The findings in this study provide references for alleviating rutting problems and improving the lifespan of asphalt pavement.

Keywords: composite high-modulus agent; rheological property; dynamic modulus; rutting resistance performance; shear resistance performance

Citation: Zhang, H.; Yang, X.; Li, Y.; Fu, Q.; Rui, H. Laboratory Evaluation of Dynamic Characteristics of a New High-Modulus Asphalt Mixture. *Sustainability* **2022**, *14*, 11838. https://doi.org/10.3390/su141911838

Academic Editors: Chaohui Wang, Dawei Wang, Kai Liu and Qian Chen

Received: 25 July 2022
Accepted: 16 September 2022
Published: 20 September 2022

1. Introduction

High-modulus asphalt concrete (HMAC) was firstly proposed by French engineers in the early 1980s for resisting permanent deformation [1]. Considering the excellent performance in alleviating the rutting problem and reducing the maximum thickness of the asphalt layer, HMAC, also known as Enrobé a Module Élevé (EME), was widely applied in both the base course and wearing course [2,3]. According to the NFP-140 standard of France [4], HMAC is generally required to have a dynamic modulus larger than 14,000 MPa at a certain testing condition (15 °C and 10 Hz). The high-modulus asphalt binder (HMAB) is an essential component for preparing the HMAC, which improves the ability of asphalt pavement to absorb stress caused by road traffic. Therefore, the development of HMAB should be an important role in the production of HMAC.

In general, HMAB is categorized into three groups, including hard-grade asphalt binder, lake/rock modified asphalt binder, and polyolefin-modified asphalt binder [5–8]. The hard-grade asphalt binder refers to those neat asphalt binders with a penetration of

10–25 (0.1 mm) at 25 °C and a softening point of 55–78 °C, which is mainly used in European countries. The second group is to use lake or rock asphalt binder as modifiers for base asphalt binder. Trinidad lake asphalt (TLA) is a common asphalt binder that is available in South America. Due to the low penetration value, TLA is not appropriate to be directly used as a binding material. Indeed, it is typically mixed with the base binder to meet specific requirements. Rock asphalt is another typical natural asphalt binder containing relatively stable physical and chemical components. Compared with the first two groups, the polyolefin-modified binder has a stronger ability in resisting thermal cracking and fatigue [9]. Considering the scarcity of the hard-grade asphalt in China and the construction convenience, the polyolefin-modified asphalt binder is the preferred selection to prepare HMAC, especially in high-temperature areas, to increase the rutting resistance of pavement.

Asphalt pavements in China have been designed for a 20-year service period. However, with the rapid development of traffic volume, in particular truck traffic, rutting has become one of the most serious problems threatening the service quality and lifespan of pavement [10]. When subjected to both heavy traffic loadings and high-temperature conditions, traditional asphalt pavements usually suffer severe deterioration in the early stage of the service period. Dense-graded asphalt concrete (DGAC) is commonly used as pavement material in most areas of China. Meanwhile, as the traffic volume increases dramatically, DGAC needs frequent maintenance to ensure pavement performance. Furthermore, many newly constructed asphalt pavements suffer rutting problems with the opening to traffic, which needs high maintenance costs. One solution to increase the life of pavement is to use material with a high modulus in its asphalt base layer. The effectiveness of this layer as a part of pavement is closely dependent on how well its material is designed. Due to the excellent anti-rutting performance, HMAC is regarded as a promising method to overcome the current problem [11–13]. However, considering the limitation of petroleum resources and production methods, the French HMAC design methodology cannot be directly used in China. Therefore, taking full advantage of the French design methodology based on the Chinese pavement design methodology is necessary to alleviate the rutting distress.

For assessing the rutting behavior of HMAC, various test methods have been adopted by researchers [14]. According to the Superpave standard, $G^*/sin\delta$ was used to evaluate the rutting potential of asphalt binder, which could be obtained by performing tests on the dynamic shear rheometer (DSR) [15,16]. Higher rutting resistance can be achieved when increasing the complex shear modulus or increasing its relative storage modulus [17]. In recent years, the multiple stress creep and recovery (MSCR) test has been recognized as a better method for evaluating the rheological property of modified asphalt binders, where the nonrecoverable creep compliance (J_{nr}) and the recovery percent (R) are two main parameters [18,19]. For the asphalt mixtures, the wheel-tracking test is the most common method to simulate the effect of tire loading on the pavement. Moreover, the Hamburg wheel tracking test [20–23] and asphalt pavement analyzer [24] are also proposed to quantitatively assess the rutting performance of asphalt mixtures. Compared with the static permanent deformation test, researchers have demonstrated that the dynamic modulus test correlated well with the actual pavement rutting behavior [25–27]. The dynamic modulus is the key index of the HMAC design methodology, and it is obtained by monitoring the response of φ 100 mm × 150 mm cylindrical specimens to dynamic compressive–shear loadings [28]. Nowadays, the published investigations on the dynamic modulus are mainly concentrated on test configurations, conditions [29], and the establishment of the dynamic modulus master curve [30,31]. It should be noticed that the shear failure, namely the lateral movement, is a primary cause of permanent deformation. However, the existing rutting tests are not able to represent the shear properties of asphalt mixtures [32,33]. Thus, conducting the shear tests is necessary to replenish the routine rutting test of asphalt mixture in the laboratory. The uniaxial penetration test (UPT) is usually applied to evaluate the shear performance of asphalt mixture at high-temperature conditions [34,35]. The Superpave shear tester (SST) is also broadly accepted by many countries. By applying different loading modes, the tests include the repeated shear test at a constant height (RSCH) and frequency sweep test at a constant height (FSCH) [36–38].

In this paper, a new high-modulus asphalt mixture was prepared, and a variety of tests were conducted to evaluate its dynamic characteristics that were rarely reported in previous researches. The high-temperature stability of HMAC was evaluated by conducting the rheological property test, dynamic modulus test, wheel tracking test, and shear resistance test. The hard-grade asphalt concrete representing the France high-modulus asphalt concrete and SBS-modified asphalt mixture widely applied in road engineering were also prepared and tested as the control group. The research results may provide some useful suggestions for asphalt mixture design to improve the rutting resistance and reduce the unnecessary cost of production, construction, and maintenance. In addition, this paper lays the experimental foundation for promoting and optimizing the new high-modulus asphalt mixture to ensure the high-temperature stability of asphalt pavement.

2. Materials and Specimen Preparation

2.1. Materials

2.1.1. Asphalt Binder

Base asphalt binder (70#, 20#) and SBS-modified asphalt binder were used in this study. The main technical indicators of asphalt binders were tested according to JTG E20-2011 (Standard Test Methods of Bitumen and Bituminous Mixtures for Highway Engineering) in China, as listed in Table 1.

Table 1. Technical indicators of base and SBS-modified asphalt binders.

Indicators	Unit	70#		20#		SBS	
		Test Value	Standard Value	Test Value	Standard Value	Test Value	Standard Value
Penetration (25 °C, 100 g, 5 s)		65.2	60–80	16.8	10–20	59.1	60–80
Softening point	°C	55.6	≥43	69.6	≥60	80.6	≥55
Ductility (5 cm/min, 15 °C)	cm	>100	≥40	10.6	/	>100	/
Relative density (15 °C)	g/cm^3	1.033	/	1.042	/	1.039	/
		After Rotating Thin-Film Oven Test (RTFOT)					
Mass loss on heating	%wt.	0.32	≤0.8	0.03	≤0.6	0.10	≤1.0
Retained penetration after RTFOT	%	78.5	≥54	85.0	≥65	84.1	≥60
Retained ductility after RTFOT	cm	59.2	≥15	4.6	/	21.3	≥20

Note: "/" means no specification in the standard.

2.1.2. Composite High-Modulus Agent

A composite high-modulus agent (CHMA) was selected as the modifier in this study, as illustrated in Figure 1. It is dark brown particles synthesized by natural rock asphalt, nanopolymer material, and stabilizer. It requires storage in a moisture-proof condition. Natural rock asphalt has good compatibility with petroleum asphalt and is less prone to aging and corrosion. Its interaction with nanopolymer material leads to the formation of an organic compound that plays an important role in enhancing the high-temperature performance of asphalt mixture. Silicon-like stabilizer ensures the storage stability of CHMA during transportation and construction. The basic properties of CHMA are presented in Table 2.

Figure 1. CHMA modifier.

Table 2. Basic properties of CHMA.

Index	Test Value	Standard Value	Test Standard
Appearance	Powder	/	Visual inspection
Color	Brown	Brown or black	Visual inspection
Density (g/cm^3)	1.40	/	T0603-2011
Flash point (°C)	307	/	T0611-2011
Water content (%)	0.82	≤2.0	T0612-2011
Natural rock bitumen content (%)	51	/	T0735-2011
Mass loss on heating after thin-film oven test (%)	0.5	≤1.0	T0609-2011

Note: "/" means no specification in the standard.

2.1.3. Aggregate Gradation

EME of France prefers continuous gradation with a smooth curve in the logarithmic coordinate system. In French standards, the passing rates of 0.063 mm, 2 mm, 4 mm, and 6 mm sieves are specified, respectively. The maximum size (D) of aggregate indicates a 100% passing rate at the sieve with 2D size, 98~100% passing rate at sieve with 1.4D size, and 85~98% passing rate at the sieve with D size. As shown in Table 3, the requirements on passing rate of key sieve size of AC-20 gradation and EME-14 gradation are very close. Hence, AC-20 asphalt mixture has been widely applied in the construction of high-modulus asphalt pavement in China. In this study, the gradation curve of AC-20 was designed as shown in Figure 2, which also met the requirements of EME14.

Table 3. Requirements on passing rate of AC-20 and EME-14.

AC-20	Sieve size (mm)	26.5	19	16	13.2	9.5	4.75	2.36	1.18	0.6	0.075
	Passing rate (%)	100	90~100	78~92	62~80	50~72	26~56	16~44	12~33	8~24	3~7
EME-14	Sieve size (mm)	28 (2D)	19.6 (1.4D)	14 (D)		6	4	2		/	0.063
	Passing rate (%)	100	98~100	85~98		50~70	40~60	25~38		/	5.4~7.7

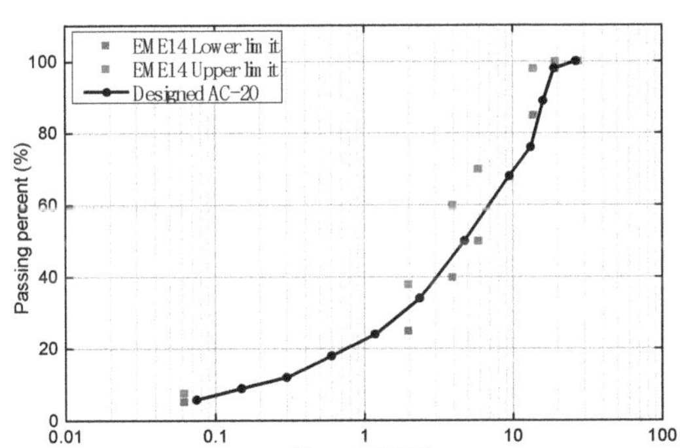

Figure 2. Aggregate gradation curve of AC-20.

2.2. Specimen Preparation

2.2.1. CHMA-Modified Asphalt Binder

The optimal content of CHMA was determined as 22% of the asphalt binder by mass according to previous research results [39]. The preparation process is briefly introduced here. First, the 70# asphalt binder was heated to 135 °C for a few hours to make it fully melted. Then, the CHMA modifier was added to the melted asphalt binder and manually mixed with a glass rod at 175 °C. Finally, the mixture was stirred for 1 h using the high-speed shear mixer at a speed of 3000 r/min to make the CHMA modifier uniformly dispersed in the base asphalt binder.

2.2.2. Asphalt Mixture Specimen

The CHMA modifier was added to the asphalt mixture through the dry process. Base asphalt mixture and SBS-modified asphalt mixture were prepared as the control group. The optimal asphalt content was 4.6% for CHMA-modified asphalt mixture, 5.2% for 20# asphalt mixture, and 5.0% for the SBS-modified asphalt mixture according to the Marshall test results. The preparation process of CHMA-modified asphalt mixture was as follows. First, the aggregates were heated to 180 °C and mixed in the pot for 30 s. The CHMA modifier was then added and mixed with the aggregate for 180 s. Next, the melted asphalt was poured into the mixing pot and mixed for 90 s. After that, the heated mineral powder was added and mixed for another 90 s. Finally, the CHMA-modified asphalt mixture was obtained, which was then transferred to the rotary compactor to prepare the cylindrical specimens with a radius of 15 cm and a height of 12 cm. It was then cut into two specimens with a height of 5 cm for FSCH and RSCH tests, as shown in Figure 3.

(**a**) (**b**)

Figure 3. The preparation process of asphalt mixture specimen: (**a**) After rotary compaction; (**b**) After cutting.

3. Methods

3.1. *Rheological Property Tests*

The frequency/temperature sweep tests were conducted to evaluate the viscoelastic properties of the CHMA-modified asphalt binders before and after the aging process using the dynamic shear rheometer (DSR) in accordance with T0628-2011 in China. For the CHMA-modified asphalt binders before and after RTFOT aged, a 25 mm parallel plate with a gap of 1 mm was used. The temperature sweep tests were performed in the range of 46–70 °C with an interval of 6 °C. The loading frequency was 10 rad/s. For the CHMA-modified asphalt binders after PAV aging, an 8 mm parallel plate with a gap of 2 mm was used. The loading frequency was 10 rad/s. Complex modulus (G*), phase angle (δ), and rutting factor (G*/sinδ) were determined from the testing data.

3.2. *Dynamic Modulus Test*

Dynamic modulus is an important index of asphalt mixture, which greatly affects the performance of asphalt pavement when subjected to dynamic vehicle loadings during the service period. Generally speaking, the asphalt mixture with a higher dynamic modulus has better rutting resistance, owing to its excellent elastic property. The dynamic modulus test was conducted under various conditions according to T0738-2011 in China. The temperature was set as 0 °C, 15 °C, 30 °C, 45 °C, and 60 °C, respectively. The loading frequency was set as 0.1 Hz, 0.5 Hz, 1 Hz, 5 Hz, 10 Hz, 20 Hz, and 25 Hz, respectively. The cylindrical specimen was loaded with the compressive force on the top surface. Two extensometers were symmetrically installed on the middle of two sides of the cylinder specimen to collect deformation information. The dynamic modulus was calculated according to Formulas (1)–(3).

$$|E^*| = \frac{\sigma_0}{\varepsilon_0} \tag{1}$$

$$\sigma_0 = \frac{P_i}{A} \tag{2}$$

$$\varepsilon_0 = \frac{\Delta_i}{l_0} \tag{3}$$

where E^* represents the dynamic modulus, MPa; σ_0 represents the axis stress amplitude, MPa; ε_0 represents the axial strain amplitude, mm/mm; P_i represents the average amplitude of compressive force during the last five loading cycles, N; A represents the radial cross-sectional area of the specimen, mm^2; Δ_i represents the average amplitude of recoverable axial deformation, mm; l_0 represents the measuring length of the extensometer, mm.

3.3. Wheel Tracking Test

Asphalt concrete (AC) consists of coarse aggregates, fine aggregates, mineral fillers, asphalt binders, and air voids. AC exhibits more viscous-like characteristics under the action of vehicle loadings, especially at high-temperature conditions, which is prone to cause rutting phenomenon. In order to evaluate the high-temperature performance of asphalt mixtures, the wheel tracking test was conducted according to T0719 (JTJE20-2011). The square slab specimens (300 mm × 300 mm× 50 mm) were fabricated and loaded with a special solid rubber tire whose contacting pressure was 0.7 MPa in a constant temperature chamber at 60 °C. The tire ran on the slab for 1 h at the speed of 42 ± 1 cycles/min during the test process. The variation of rutting depth with running time was measured and recorded, and the dynamic stability (DS) could be expressed as Formula (4).

$$DS = \frac{15 \times 42}{d_{60} - d_{45}} = \frac{630}{d_{60} - d_{45}} \tag{4}$$

where d_{45} and d_{60} are the rutting depth at 45 min and 60 min, respectively.

3.4. Frequency Sweep Test at a Constant Height

The FSCH test was conducted on the specimen to investigate the influence of vehicle speed on the rutting behavior of asphalt mixtures. The operating details were specified in AASHTO T 320. The strain control mode was adopted, as illustrated in Figure 4. The horizontal strain was selected as a sine wave with an amplitude of 50 µm. The loading frequency was determined as 10 Hz, 5 Hz, 2 Hz, 1 Hz, 0.5 Hz, 0.2 Hz, 0.1 Hz, 0.05 Hz, 0.02 Hz, and 0.01 Hz. The loading cycles varied with the loading frequency, namely 50 loading cycles for 10 Hz and 5 Hz, 20 loading cycles for 2 Hz and 1 Hz, 7 loading cycles for 0.5 Hz, 0.2 Hz and 0.1 Hz, and 4 loading cycles for 0.05 Hz, 0.02 Hz, and 0.01 Hz. Test temperature was set as 60 °C. During the FSCH test, the vertical force was provided to keep the specimen height constant, and the horizontal loads with different frequencies were provided to simulate the effect of different driving speeds on the asphalt pavement. The stress and strain of the specimen were recorded during the test period.

The complex shear modulus G^* at different frequencies can be calculated using the stress and strain data, as shown in Formula (5).

$$G^* = \frac{\tau_0}{\gamma_0} \tag{5}$$

where G^* represents the shear modulus; τ_0 represents the peak value of dynamic shear stress; γ_0 represents the recoverable deformation.

Shear stress and strain at time t can be calculated according to Formulas (6) and (7).

$$\tau_t = \tau_0 \sin(\omega t) \tag{6}$$

where τ_t represents the shear stress at time t; ω represents the angular velocity; t represents time.

$$\gamma_t = \gamma_0 \sin(\omega t - \varphi) \quad (7)$$

where γ_t represents the shear strain at time t; γ_0 represents the recoverable strain; φ represents phase angle.

The phase angle is used to characterize the viscoelastic properties of materials. It is 0 for pure elastic materials, while it is 90 for pure viscous materials. The complex shear modulus is closely related to the deformation ability of asphalt concrete. Therefore, the phase angle and complex shear modulus were adopted to evaluate the resistance of asphalt concrete to shear deformation.

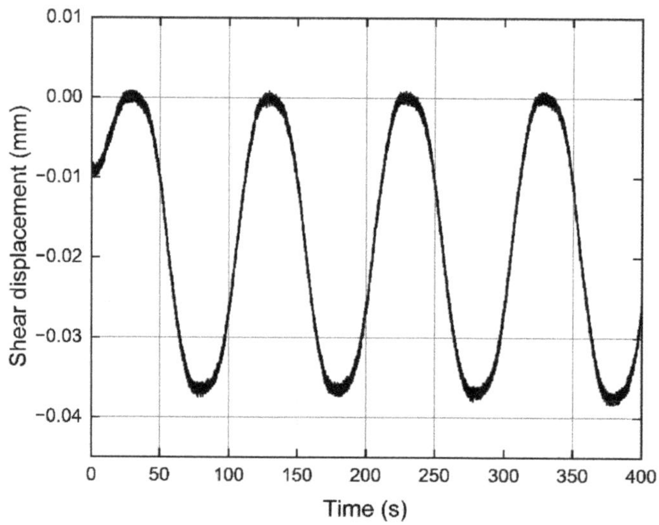

Figure 4. Loading mode of FSCH test.

3.5. Repeated Shear Test at a Constant Height

The RSCH test is a shear dynamic modulus test for measuring the stress–strain relationship of the specimen under shear loadings at a constant shear strain in accordance with AASHTO T 320.

During the test process, the specimen height was kept constant, and a half sine wave of 69 ± 5 kPa was applied to the specimen for 0.1 s and then resting for 0.6 s in each loading cycle, as shown in Figure 5. The test temperature was set as 60 °C. The variation of strain with the number of repeated loadings was recorded, which simulated the response of asphalt pavement to repeated vehicle loadings. The regression equation was obtained as Formula (8) based on the cumulative strain curve. The antishear performance of asphalt mixture at high temperatures could be characterized by the shear slope.

$$\varepsilon(\mathrm{N}) = \varepsilon(\mathrm{N} = 1) + \mathrm{S} \times \mathrm{N} \quad (8)$$

where $\varepsilon(\mathrm{N})$ represents the permanent strain after the Nth loading cycle; $\varepsilon(\mathrm{N} = 1)$ represents the permanent strain after the first loading cycle; S represents the shear slope; N represents the number of loading cycles.

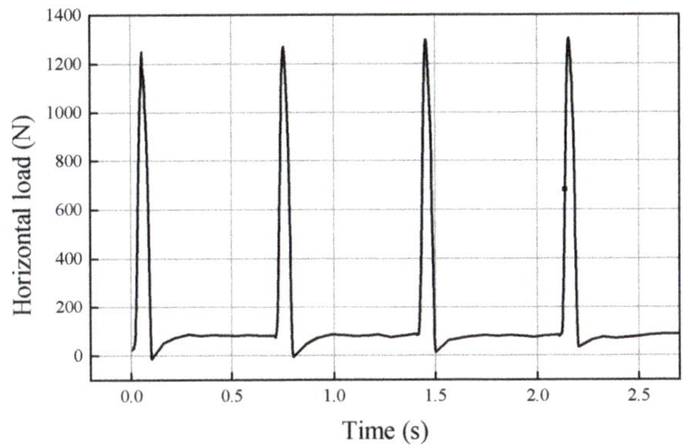

Figure 5. Loading mode of RSCH test.

4. Results and Discussions

4.1. Rheological Propery of Asphalt Binders

The rheological property of the asphalt binders was investigated by the frequency sweep test and the temperature sweep test, respectively. The variation of the complex shear modulus G*, phase angle δ, and rutting factor G*/sinδ with the frequency was calculated and presented in Figure 6a–c, respectively. G*/sinδ is used to reflect the ability of the asphalt binder to resist deformation. The higher value of G* indicates smaller deformation and better performance at a high-temperature condition. With the increase of frequency, G* went up quickly, while δ dropped down gradually. It should be noted that the changing rates of the 20# asphalt binder and CHMA-modified asphalt binder was obviously higher than that of SBS-modified asphalt binder. CHMA-modified asphalt binder was close to 20# asphalt binder in G* value, and their rutting factor showed the same tendency, demonstrating its strong ability to resist deformation.

The rutting factors of neat, RTFOT aged, and PAV aged asphalt binders are illustrated in Figure 7. The temperature had a significant effect on the properties of the asphalt binders. With the temperature increasing, the complex modulus declined, and the phase angle increased. Meanwhile, the rutting factor also decreased, indicating that the viscous characteristic of asphalt binders gradually increased, negatively affecting its high-temperature stability. For the RTFOT and PAV aged asphalt binders, the variation of rutting factor with temperature also presented a similar regular pattern. In addition, the value of rutting factor kept growing with the aggravation of aging. This can be explained by the increase of elastic response and the decrease of viscous response due to the hardening of the asphalt binder during the aging process.

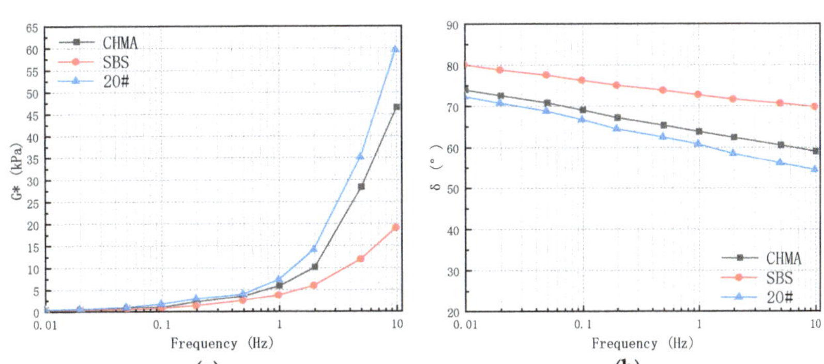

Figure 6. *Cont.*

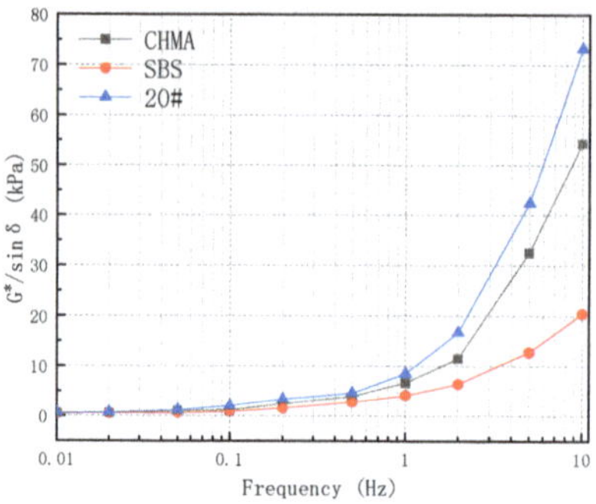

(c)

Figure 6. Frequency sweep test results: (**a**) complex shear modulus; (**b**) phase angle; (**c**) rutting factor.

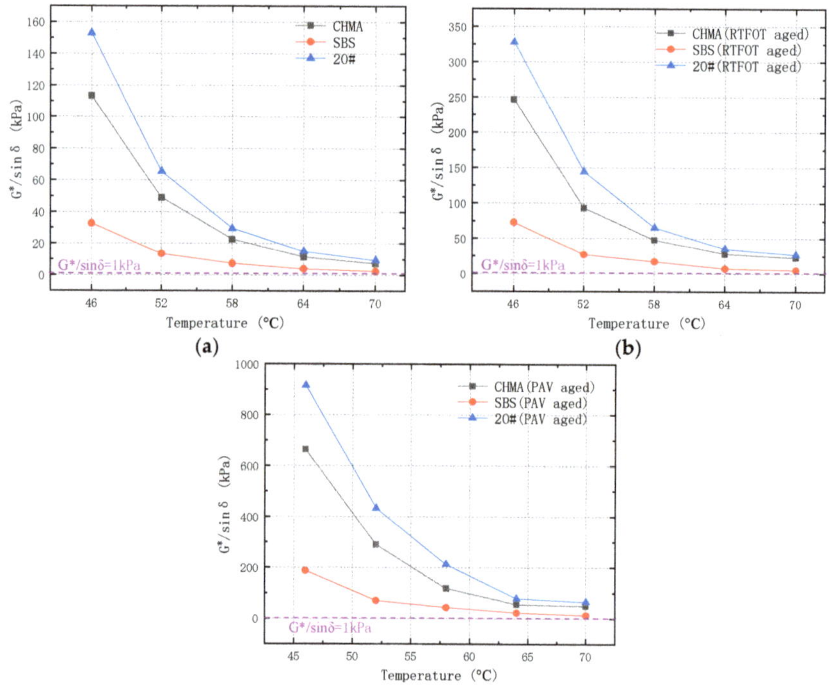

(c)

Figure 7. Temperature sweep test results: (**a**) neat asphalt binders; (**b**) RTFOT aged asphalt binders; (**c**) PAV aged asphalt binders.

4.2. Dynamic Mechanical Performance

4.2.1. Dynamic Modulus

The dynamic modulus and phase angle of the asphalt mixtures at various testing temperatures are shown in Figure 8. The dynamic modulus increased with the raising frequency and decreased with the elevating temperature. That is to say, the lower vehicle speed and higher temperature led to a lower dynamic modulus, which was consistent with the phenomenon that asphalt pavement was prone to suffer rutting distress when subjected to low-speed vehicle loadings, especially at high temperatures. This indicated that the high-modulus asphalt mixtures could effectively alleviate rutting at high temperatures and low vehicle speed conditions. Among the three types of asphalt mixtures, AC-20(20#) had the highest dynamic modulus, and AC-20(CHMA) was slightly lower than AC-20(20#), while AC-20(SBS) presented the worst performance. At 15 °C and 10 Hz conditions (Figure 8b), the dynamic modulus of AC-20(20#), AC-20(CHMA), and AC-20(SBS) were 22,915 Mpa, 19,568 Mpa, and 12,497 Mpa, respectively. The first two types of mixtures exceeded 14,000 Mpa, which met the specification requirement.

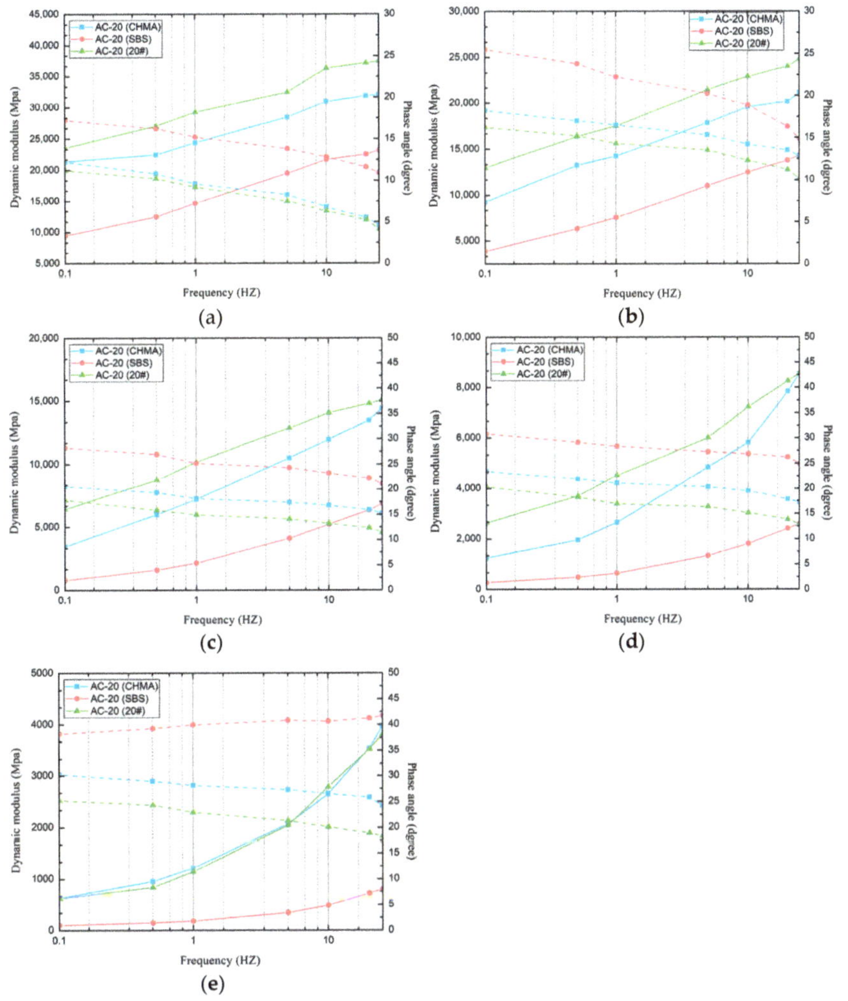

Figure 8. Dynamic modulus test results of various mixtures: (**a**) 0 °C; (**b**) 15 °C; (**c**) 30 °C; (**d**) 45 °C; (**e**) 60 °C.

The phase angle mainly showed decreasing tendency with the increase of frequency. This could be attributed to the increase in elastic response of the asphalt binder in the high-frequency range. It should be noted that, in Figure 8e, the phase angle of AC-20(SBS) exhibited different characteristics compared with the other two types of asphalt mixtures, which went up gradually with the increase in the frequency. It could be explained by the stress-dependent behaviors of the aggregate skeleton and the viscosity changes. When the temperature reached a certain value, the asphalt binder behaved as a viscous material, which affected the performance of mixtures significantly. Therefore, the phase angle of

AC-20(SBS) was the largest and exhibited a fluctuating tendency. The other two types of asphalt mixtures did not show obvious viscous characteristics, implying stronger resistance to rutting at high temperatures.

4.2.2. Dynamic Modulus Master Curve

Due to the limitation of the testing device, time, and other causes, the results of the dynamic modulus could be tested only in a certain condition. In order to predict the dynamic modulus under various temperatures, frequencies, and extreme conditions, it is necessary to establish the dynamic modulus master curve. First, the sigmoidal function was selected as the fitting model to fit the test results at different temperature conditions. Then, for a certain reference temperature, the shift factors at various test temperatures were calculated based on the principle of time–temperature equivalence. Finally, the dynamic modulus curves at different temperatures and loading frequencies were shifted to the reference temperature, and superimposed into a smooth curve, namely the master curve [40].

Figure 9a shows the scatter diagram of dynamic modulus of CHMA-modified asphalt mixture under different frequencies after being shifted from other test temperatures to 30 °C. The sigmoidal function, as shown in Formula (9), was used to fit the data in Figure 9a using the simple surface climbing method. The values of the fitting coefficients were obtained, as shown in Table 4. The dynamic modulus at different temperatures and the corresponding master curves of the control groups are depicted in Figures 10 and 11. Similarly, their fitting coefficients were also obtained, as shown in Table 4.

$$log(E^*) = \delta + \frac{\varphi - \delta}{1 + e^{\beta + \gamma \cdot log\omega_r}} \tag{9}$$

where E^* is the dynamic modulus; ω_r is the reduced frequency at corresponding temperatures; δ is the logarithm of the minimum value of dynamic modulus; φ, β, and γ are the fitting coefficients.

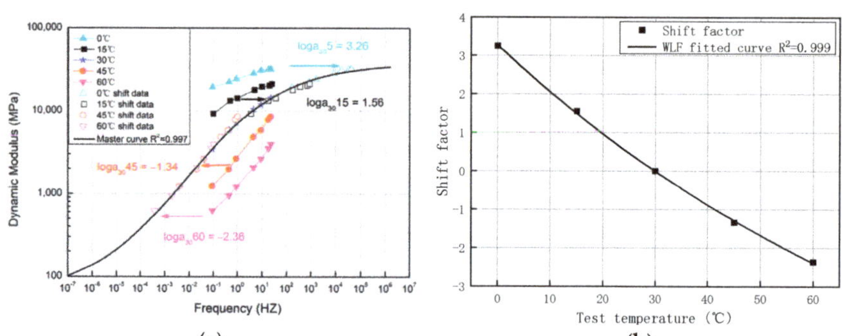

Figure 9. Master curves of dynamic modulus of AC-20(CHMA): (**a**) master curve of dynamic modulus; (**b**) shift factor at different temperatures.

Table 4. Fitting coefficients for dynamic modulus master curve.

Fitting Coefficient		δ	φ	β	γ	C_1	C_2
Estimated value	AC-20(CHMA)	1.772	4.574	−1.051	−0.493	17.9	193.5
	AC-20(SBS)	1.120	3.269	−0.864	−0.532	23.3	248.6
	AC-20(20#)	1.793	4.627	−1.266	−0.395	362.7	3055.6

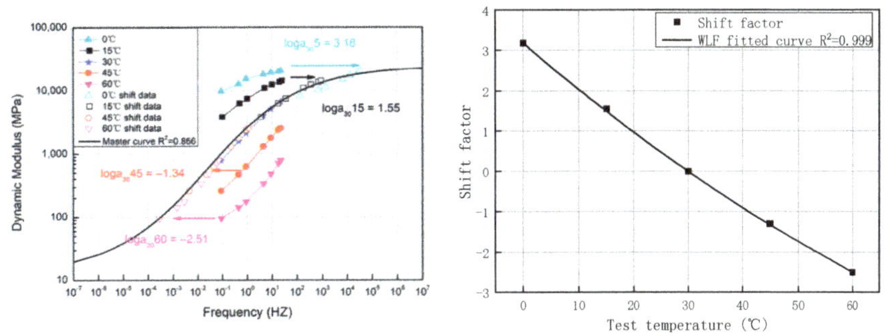

(a) (b)

Figure 10. Master curves of dynamic modulus of AC-20(SBS): (**a**) master curve of dynamic modulus; (**b**) shift factor at different temperatures.

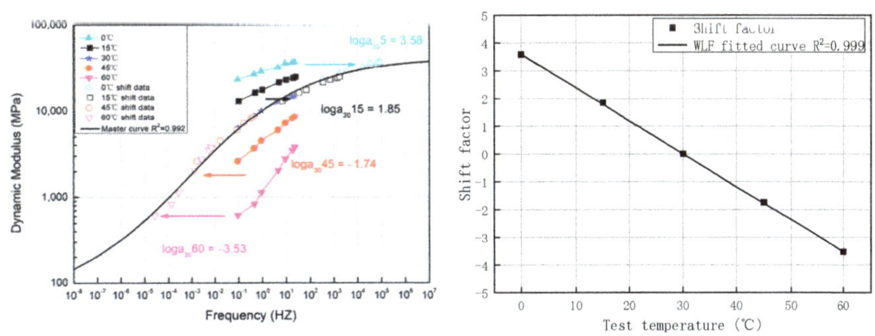

(a) (b)

Figure 11. Master curves of dynamic modulus of AC-20(20#): (**a**) master curve of dynamic modulus; (**b**) shift factor at different temperatures.

Furthermore, the WLF (Williams–Landel–Ferry) empirical equation, as shown in Formula (10), was used to fit the shift factors at different temperatures. It can be seen from Figure 9b that the correlation coefficient of the WLF fitted curve was above 0.99. It implies that the shift factor could be well fitted by the WLF empirical equation, and it is feasible to analyze the interlayer dynamic modulus using the time–temperature equivalent principle. The values of C_1 and C_2 were obtained in Table 4.

$$\lg\alpha_{30}(T) = \frac{-C_1(T-30)}{C_2+(T-30)} \tag{10}$$

where T is the test temperature; $\alpha_{30}(T)$ is the shift factor of test temperature T relative to the reference temperature (30 °C); C_1 and C_2 are the fitting coefficients.

The dynamic modulus curve at any other temperatures can be obtained by constructing the master curve combined with the shift factor, and the changing trend of dynamic modulus in a wider range of loading frequencies can also be known. This lays the foundation for the mechanical analysis of asphalt pavement with a high modulus.

The logarithmic value of the dynamic modulus approaches the minimum value (δ) at the tail end of the master curve, reflecting the mechanical property of asphalt mixture at high-temperature and low-frequency conditions. It approaches the maximum value (φ) at the head end of the master curve, reflecting the mechanical property of asphalt mixture at low-temperature and high-frequency conditions. Hence, the two ends of the master curve can be used to explain the mechanical properties of asphalt mixture under extreme conditions. From Figures 9–11 and Table 4, it can be found that the dynamic modulus gradually increased with increasing frequency. Under the testing conditions, AC-20(20#) exhibited the best mechanical performance regardless of whether it was at the high-temperature, low-frequency condition or the low-temperature, high-frequency condition. AC-20(CHMA) was only slightly weaker than AC-20(20#) in mechanical strength, and obviously superior to AC-20 (SBS), indicating that it had potentially good application in constructing high modulus asphalt pavement.

4.3. Rutting Resistance Performance

Figure 12 illustrates the results of the wheel tracking test. The dynamic stability of AC-20(SBS), AC-20(CHMA), and AC-20(20#) was 6698 times/mm, 8094 times/mm, and 10,624 times/mm, respectively. Although they all met the requirement of JTG F40-2004, which requires more than 3000 times/mm, they showed different resistance abilities to the rutting deformation. Among these mixtures, AC-20(20#) had the best rutting performance, whose dynamic stability was 1.31 times and 1.59 times that of the AC-20(CHMA) and AC-20(SBS), respectively. In addition, the final deformations of three asphalt mixtures were 1.957 mm, 1.425 mm, and 1.082 mm, which were consistent with the results of dynamic stability. This could be ascribed to the viscoelastic characteristic of asphalt binder. Since the higher elastic characteristic of CHMA-modified asphalt binder compared with SBS-modified asphalt binder, CHMA-modified asphalt mixture had better permanent deformation resistance than SBS-modified asphalt mixture. Advantages of CHMA-modified asphalt at high temperatures was due to the compound high-modulus modifier HRMA enriched with natural rock asphalt. Natural rock asphalt has the characteristics of a high softening point and high nitrogen content. Its addition improved the softening point and cohesion of the matrix asphalt. Therefore, the high-temperature performance of the asphalt mixture was improved.

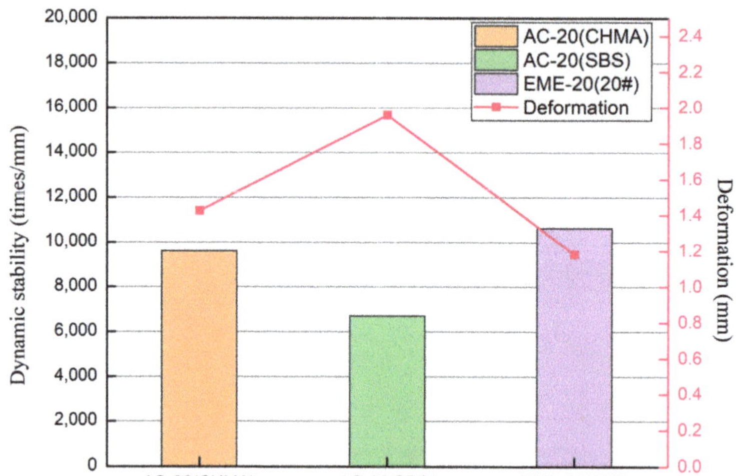

Figure 12. Rutting test results of various asphalt mixtures.

4.4. Shear Resistance Performance

4.4.1. FSCH Test Results

The FSCH test results are presented in Figure 13. In the low-frequency range (0.01–0.1 Hz), there was no obvious difference among these mixtures. With the increase in frequency, the mixtures showed different development trends of complex shear modulus. AC-20(20#) had the highest complex shear modulus and increasing rate, indicating that it was sensitive to the frequency. It should be mentioned that the complex shear modulus usually exhibited a small value in the low-frequency range, which indicated that low velocity affected the pavement performance significantly. According to the investigation in China, 80 km/h is the most probable velocity in the highway, which corresponds to 10 Hz. At the frequency of 10 Hz, the complex shear modulus of AC-20(20#), AC-20(CHMA), and AC-20(SBS) were 241.5 MPa, 185.3 Mpa, and 65.5 MPa, respectively. Thus, AC-20(20#) and AC-20(CHMA) both showed good performance according to the FSCH test results.

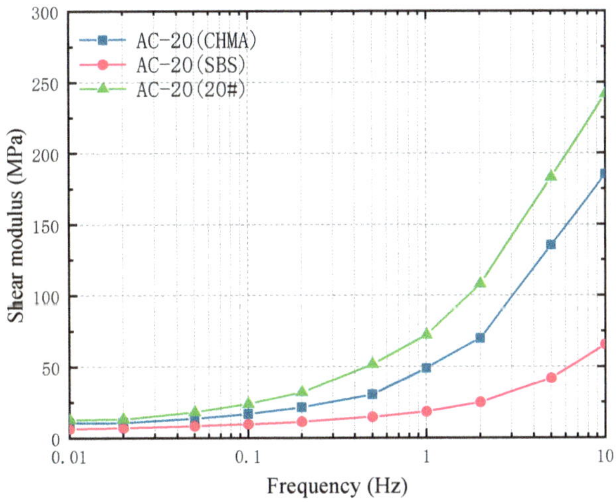

Figure 13. Complex shear modulus of various mixtures.

Figure 14 presents the variation of phase angle with the frequency in the FSCH test. The phase angle of AC-20(CHMA) was always between the other two mixtures. With the increase of frequency, the phase angle gradually increased until a peak value was reached around 1Hz, and then it experienced a downward trend. In the low-frequency range, the phase angle of AC-20(CHMA) was slightly higher than that of AC-20(20#), while AC-20(SBS) had the highest phase angle. In the high-frequency range, mixtures showed different levels. Among these mixtures, AC-20(20#) had the lowest phase angle, indicating that it was dominated by elastic characteristics.

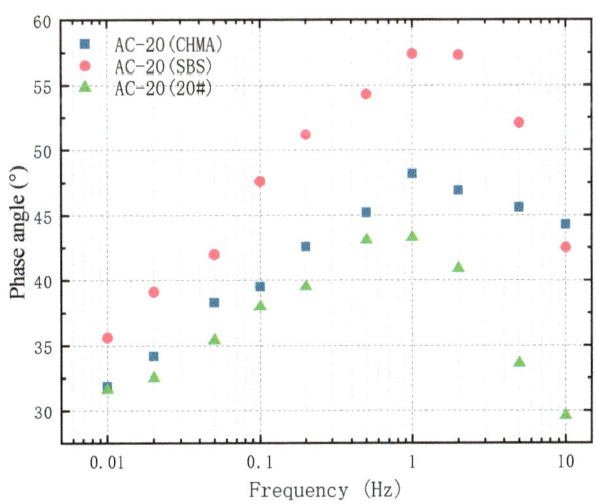

Figure 14. Phase angle of asphalt mixtures.

4.4.2. RSCH Test Results

Asphalt pavement usually suffers rutting distress due to inadequate shear resistance under repeated vehicle loadings, especially at high-temperature condition. The RSCH test was adopted to evaluate the shear resistance ability of asphalt mixture. Test results are presented in Figures 15 and 16. With the number of loading cycles increasing, the shear strain experienced a rapid increase in the primary stage, and then its growth rate slowed down gradually, finally reaching a constant growth rate. The final strain of AC-20(SBS) was the highest, followed by AC-20(CHMA) and AC-20(20#). It means that the high-modulus asphalt mixtures had a distinct advantage. The strain variation trend of AC-20(CHMA) was similar to that of AC-20(20#), and its growth rate was a little higher than AC-20(20#). Hence, AC-20(CHMA) was recognized to be excellent in resisting shear deformation. Based on the shear strain curve, the shear slopes of AC-20(SBS), AC-20(CHMA), and AC-20(20#) were calculated as 2.26×10^{-4}, 1.45×10^{-4}, and 6.89×10^{-5}, respectively. The smaller the shear slope is, the better the shear resistance ability is. AC-20(20#) had the smallest shear slope, while AC-20(SBS) had the largest shear slope. Although the shear slope of AC-20(CHMA) was higher than AC-20(20#), it was obviously lower than AC-20(SBS), demonstrating that the CHMA-modified asphalt binder contributed a lot to the antirutting performance of asphalt mixture.

AC-20(CHMA) has excellent high-temperature shear resistance, which is due to the modification of matrix asphalt by compound high-modulus modifier HRMA. The modifier contains a certain proportion of natural rock asphalt. Natural rock asphalt is a kind of natural solid asphalt with relatively large molecular weight. Its molecular weight is about three times that of ordinary asphalt, and the content of carbon, hydrogen, sulfur, oxygen, and nitrogen is relatively high. Almost every asphalt macromolecule contains the above impurities. The polar functional groups of each element make it have strong adsorption capacity on the rock surface. In other words, the addition of HRMA modifier improves the cohesion and shear deformation resistance of asphalt mortar.

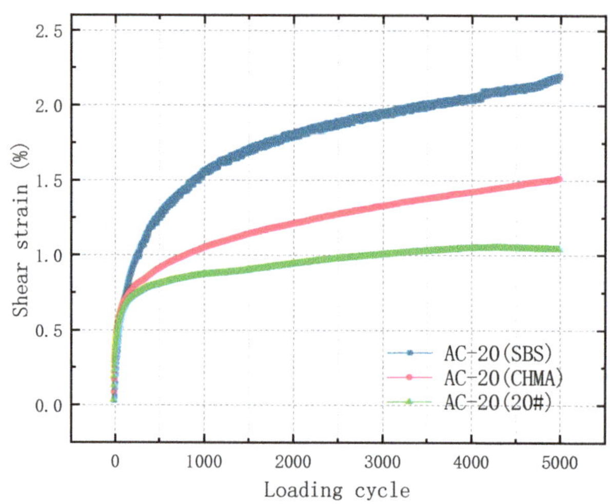

Figure 15. Variation of shear strain with loading cycles.

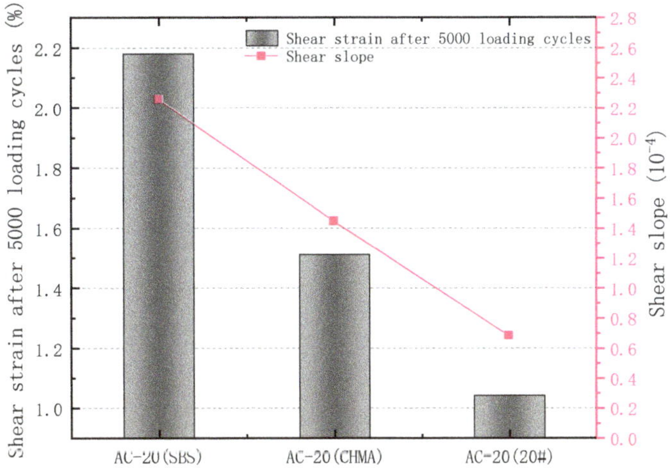

Figure 16. Shear slope and strain after 5000 loading cycles.

5. Conclusions

In this paper, a new high-modulus asphalt mixture, namely the CHMA-modified asphalt mixture, was prepared and investigated, aiming to improve the antirutting performance of asphalt pavement. The dynamic characteristics of CHMA-modified asphalt binder and mixture were investigated by the frequency/temperature sweep test, dynamic modulus test, wheel tracking test, FSCH test, and RSCH test to analyze its deformation-resisting ability. The main conclusions were obtained as follows:

(1) The rheological property of the CHMA-modified asphalt binder was similar to that of low-graded asphalt binder, demonstrating that it has a strong ability to resist deformation.
(2) The dynamic modulus of AC-20(CHMA) was 19,568 MPa at 15 °C and 10 Hz conditions, higher than 14,000 MPa, meeting the requirement for the high-modulus asphalt mixture. It had potentially good application in alleviating the rutting problems.
(3) The dynamic stability of AC-20(CHMA) was 8094 times/mm, lower than that of AC-20(20#), but obviously higher than that of AC-20(SBS). Meanwhile, it was much higher than the specification requirement in China.
(4) The asphalt mixture had a low complex shear modulus in the low-frequency range, implying that low velocity affected the pavement performance significantly. AC-20(20#) and AC-20(CHMA) both showed strong shear resistance according to the FSCH test results.
(5) Under the repeated shear loadings, the growth rate of shear strain increased rapidly in the primary stage, and then slowed down gradually, finally reaching a constant growth rate. The shear slope of AC-20(CHMA) was higher than AC-20(20#), and obviously lower than AC-20(SBS).

Author Contributions: Conceptualization, H.Z. and X.Y.; methodology, H.Z.; software, Y.L.; validation, H.R., Y.L. and Q.F.; formal analysis, X.Y. and H.Z.; investigation, X.Y.; resources, H.Z. and Y.L.; data curation, Y.L. and Q.F.; writing—original draft preparation, Y.L.; writing—review and editing, Q.F. and Y.L.; visualization, X.Y.; supervision, X.Y.; project administration, H.Z.; funding acquisition, H.Z. All authors have read and agreed to the published version of the manuscript.

Funding: This research was funded by the Science and Technology Department of Henan Province (no. 202102310263, no. 212102310986) and Science and Technology Planning Project of Department of Housing and Urban-Rural Development of Henan Province (HNJS-2020-K36).

Institutional Review Board Statement: Not applicable.

Informed Consent Statement: Not applicable.

Data Availability Statement: The data presented in this study are available on request from the corresponding author.

Acknowledgments: The authors acknowledge the support given by Bowei Sun Checking and Proofreading this article.

Conflicts of Interest: The authors declare no conflict of interest.

References

1. Nunn, M.E.; Smith, T. *Road Trials of High Modulus Base for Heavily Trafficked Roads*; Thomas Telford: Telford, UK, 1999.
2. Serfass, J.; Bense, P.; Pellevoisin, P. Properties and new developments of high modulus asphalt concrete. In Proceedings of the Eighth International Conference on Asphalt Pavements, Seattle, WA, USA, 1 January 1997.
3. Chen, Y.; Wang, H.; Xu, S.; You, Z. High modulus asphalt concrete: A state-of-the-art review. *Constr. Build. Mater.* **2020**, *237*, 117653. [CrossRef]
4. Association Francaise de Normalisation (AFNOR). *NF EN 13108-1*; Bituminous Mixtures e Material Specifications—Part 1: Asphalt Concrete. Association Francaise de Normalisation: Paris, France, 2007.
5. Wang, X.; Qiu, Y.; Xue, S.; Yang, Y.; Zheng, Y. Study on durability of high-modulus asphalt mixture based on TLA and fibre composite modification technology. *Int. J. Pavement Eng.* **2018**, *19*, 930–936. [CrossRef]
6. Oda, S.; Fernandes, J.L.; Ildefonso, J.S. Analysis of use of natural fibers and asphalt rubber binder in discontinuous asphalt mixtures. *Constr. Build. Mater.* **2012**, *26*, 13–20. [CrossRef]
7. Sefidmazgi, N.R.; Tashman, L.; Bahia, H. Internal structure characterization of asphalt mixtures for rutting performance using imaging analysis. *Road Mater. Pavement Des.* **2012**, *13*, 21–37. [CrossRef]
8. Wang, C.H.; Chen, Q.; Guo, T.T.; Li, Q. Environmental Effects and Enhancement Mechanism of Graphene/tourmaline Composites. *J. Clean. Prod.* **2020**, *262*, 121313. [CrossRef]
9. Geng, H.; Clopotel, C.S.; Bahia, H.U. Effects of high modulus asphalt binders on performance of typical asphalt pavement structures. *Constr. Build. Mater.* **2013**, *44*, 207–213. [CrossRef]
10. Chen, Q.; Wang, C.; Yu, S.; Song, Z.; Fu, H.; An, T. Low-Temperature Mechanical Properties of Polyurethane-Modified Waterborne Epoxy Resin for Pavement Coating. *Int. J. Pavement Eng.* **2022**, *23*, 1–13. [CrossRef]
11. Judycki, J.; Jaczewski, M.; Rys, D.; Pszczoła, M.; Jaskula, P.; Glinicki, A. Field investigation of low-temperature cracking and stiffness moduli on selected roads with conventional and high modulus asphalt concrete. *IOP Conf. Ser. Mater. Sci. Eng.* **2017**, *236*, 012002. [CrossRef]
12. Zhang, Z.Y.; Wang, C.H.; Zhang, L.; Wang, S.; Zeng, W. Construction and assessment technology of green road in China. *J. Chang'an Univ. (Nat. Sci. Ed.)* **2018**, *38*, 76–86. (In Chinese)
13. Zheng, M.; Li, P.; Yang, J.; Li, H.; Qiu, Y.; Zhang, Z. Fatigue character comparison between high modulus asphalt concrete and matrix asphalt concrete. *Constr. Build. Mater.* **2019**, *206*, 655–664. [CrossRef]
14. Alimohammadi, H.; Zheng, J.; Buss, A.; Schaefer, V.; Zheng, G. Rutting Performance Evaluation of Hot Mix Asphalt and Warm Mix Asphalt Mixtures by Using Dynamic Modulus, Hamburg Wheel Tracking Tests, and Viscoelastic Finite Element Simulations. In Proceedings of the International Conference on Transportation and Development 2020: Highway and Airfield Pavements, Seattle, WA, USA, 26–29 May 2020.
15. Kennedy, T.W.; Huber, G.A.; Harrigan, E.T.; Cominsky, R.J.; Hughes, C.S.; Von Quintus, H.; Moulthrop, J.S. *Superior Performing Asphalt Pavements (Superpave): The Product of the Shrp Asphalt Research Program, Strategic Highway Research Program*; National Research Council: Washington, DC, USA, 1994.
16. Laukkanen, O.V.; Soenen, H.; Pellinen, T.; Heyrman, S.; Lemoine, G. Creep-recovery behavior of bituminous binders and its relation to asphalt mixture rutting. *Mater. Struct.* **2015**, *48*, 4039–4053. [CrossRef]
17. Han, M.; Li, J.; Muhammad, Y.; Hou, D.; Zhang, F.; Yin, Y.; Duan, S. Effect of polystyrene grafted graphene nanoplatelets on the physical and chemical properties of asphalt binder. *Constr. Build. Mater.* **2018**, *174*, 108–119. [CrossRef]
18. Hajikarimi, P.; Rahi, M.; Nejad, F.M. Comparing Different Rutting Specification Parameters Using High Temperature Characteristics of Rubber-Modified Asphalt Binders. *Road Mater. Pavement Des.* **2015**, *16*, 751–766. [CrossRef]
19. Li, Y.; Hao, P.; Zhao, C.; Ling, J.; Wu, T.; Li, D.; Liu, J.; Sun, B. Anti-Rutting Performance Evaluation of Modified Asphalt Binders: A Review. *J. Traffic Transp. Eng.* **2021**, *8*, 339–355. [CrossRef]
20. Guo, R.H.; Prozzi, J.A. A statistical analysis of Hamburg wheel tracking device testing results, Road Mater. *Pavement* **2009**, *10*, 327–335. [CrossRef]
21. Rafiq, W.; Napiah, M.B.; Sutanto, M.H.; Salah Alaloul, W.; Nadia Binti Zabri, Z.; Imran Khan, M.; Ali Musarat, M. Investigation on hamburg wheel-tracking device stripping performance properties of recycled hot-mix asphalt mixtures. *Materials* **2020**, *13*, 4704. [CrossRef]
22. Sel, I.; Yildirim, Y.; Ozhan, H.B. Effect of test temperature on Hamburg wheel-tracking device testing. *J. Mater. Civil. Eng.* **2014**, *26*, 04014037. [CrossRef]

23. Tsai, B.W.; Coleri, E.; Harvey, J.T.; Monismith, C.L. Evaluation of AASHTO T 324 Hamburg-wheel track device test. *Constr. Build. Mater.* **2016**, *114*, 248–260. [CrossRef]
24. Rushing, J.F.; Garg, N. Using the asphalt pavement analyzer as a mixture performance test to select appropriate binder grades for airport pavements. *J. Transp. Eng. B-Pave.* **2017**, *143*, 04017010. [CrossRef]
25. Walubita, L.F.; Zhang, J.; Das, G.; Hu, X.; Mushota, C.; Alvarez, A.E.; Scullion, T. Hot-mix asphalt permanent deformation evaluated by Hamburg wheel tracking, dynamic modulus, and repeated load tests. *J. Transport. Res. Rec.* **2012**, *2296*, 46–56. [CrossRef]
26. Zhang, J.; Alvarez, A.E.; Lee, S.I.; Torres, A.; Walubita, L.F. Comparison of flow number, dynamic modulus, and repeated load tests for evaluation of HMA permanent deformation. *Constr. Build. Mater.* **2013**, *44*, 391–398. [CrossRef]
27. Goh, S.W.; You, Z.P.; Williams, R.C.; Li, X. Preliminary dynamic modulus criteria of HMA for field rutting of asphalt pavements: Michigan's experience. *J. Transp. Eng-ASCE* **2011**, *137*, 37–45. [CrossRef]
28. Alimohammadi, H.; Zheng, J.; Buss, A.; Schaefer, V.R.; Williams, C.; Zheng, G. Field and Simulated Rutting Behavior of Hot Mix and Warm Mix Asphalt Overlays. *Constr. Build. Mater.* **2020**, *265*, 120366. [CrossRef]
29. Kim, Y.R.; Seo, Y.; King, M.; Momen, M. Dynamic modulus testing of asphalt concrete in indirect tension mode. *Bitum. Paving Mixtures* **2004**, *1891*, 163–173. [CrossRef]
30. Pellinen, T.K.; Witczak, M.W.; Bonaquist, R.F. Asphalt mix master curve construction using sigmoidal fitting function with non-linear least squares optimization technique. In Proceedings of the 15th ASCE Engineering Mechanics Conference, New York, NY, USA, 2–5 June 2002.
31. Luo, R.; Liu, H. Improving the accuracy of dynamic modulus master curves of asphalt mixtures constructed using uniaxial compressive creep tests. *J. Mater. Civ. Eng.* **2017**, *29*, 04017032. [CrossRef]
32. Walubita, L.; Lee, S.; Zhang, J.; Faruk, A.N.; Nguyen, S.; Scullion, T. *HMA Shear Resistance, Permanent Deformation, and Rutting Tests for Texas mixes: Year-1 Report*; Technical Report FHWA/TX-13/0-6744-1; Texas A&M Transportation Institute: College Station, TX, USA, 2013.
33. Brown, E.R.; Kandhal, P.S.; Zhang, J. *Performance Testing for Hot Mix Asphalt*; NCAT Report 01-05; Auburn University: Auburn, AL, USA, 2001.
34. Chen, X.; Huang, B.; Xu, Z. Uniaxial penetration testing for shear resistance of hot-mix asphalt mixtures. *Transp. Res. Rec. J. Transp. Res. Board* **2006**, *1970*, 116–125. [CrossRef]
35. Bi, Y. Research on Anti-Shear Test Method and Parameter of Asphalt Mixture. Ph.D. Dissertation, School of Transportation Engineering, Tongji University, Shanghai, China, 2004.
36. Romero, P.; Mogawer, W. Evaluation of the Superpave shear tester using 19-mm mixtures from the Federal Highway Administration's accelerated loading facility, Technical Session on Asphalt Paving Technology 1998. In Proceedings of the 73rd Annual Meeting of the Association-of-Asphalt-Paving-Technologists, Boston, MA, USA, 16–18 March 1998; pp. 573–601.
37. Shatnawi, S.R.; Lancaster, F. *Field and Laboratory Evaluation of Superpave Level One Mix Design in California, Symposium on Progress of Superpave (Superior Performing Asphalt Pavement)—Evaluation and Implementation*; ASTM Special Technical Publication: New Orleans, LA, USA, 1996; pp. 64–79.
38. Zhang, J.N.; Xie, H.B.; Kandhal, P.S.; Powell, R.D. Field Validation of Superpave Shear Test on NCAT Test Track. In Proceedings of the Symposium on Performance of Hot Mix Asphalt (HMA), Tampa, FL, USA, 9–10 December 2003.
39. Mi, S.Z.; Li, Y.X.; Zhang, H.W. Preparation and Characterization of High Modulus Agent Modified Asphalt and Its High Modulus Mixture. *Adv. Mater. Sci. Eng.* **2022**, *2022*, 2374241. [CrossRef]
40. Alimohammadi, H.; Zheng, J.; Buss, A.; Schaefer, V.R.; Williams, C.; Zheng, G. Finite Element Viscoelastic Simulations of Rutting Behavior of Hot Mix and Warm Mix Asphalt Overlay on Flexible Pavements. *Int. J. Pavement Res. Technol.* **2021**, *14*, 708–719. [CrossRef]

sustainability

MDPI

Article

Fatigue Properties and Damage Characteristics of Polyurethane Mixtures under a Stress Control Mode

Min Sun [1], Guangzhen Qu [2,*], Litao Geng [1], Derui Hou [1] and Shuo Jing [1]

[1] School of Transportation Engineering, Shandong Jianzhu University, Jinan 250101, China
[2] School of Civil Engineering and Architecture, University of Jinan, Jinan 250022, China
* Correspondence: quguangzhen@126.com

Abstract: A polyurethane mixture (PUM) is an energy-saving and emission-reducing pavement material with excellent temperature stability; however, the fatigue properties and fatigue damage models of PUM still require further research. Therefore, four-point bending static load tests, fatigue tests, and digital speckle correlation method (DSCM) tests with different load levels of PUM and styrene butadiene styrene (SBS)-modified asphalt mixture (SMA) were carried out. The fatigue life, stiffness, midspan deflection, and maximum tensile strain were obtained and compared. The fatigue damage factor calculation method of PUM based on stiffness degradation was proposed, and the fatigue damage function of PUM at different load levels was fitted. The results show that the fatigue life of PUM was much larger than that of SMA, and the static loading failure and fatigue failure modes of PUM were both brittle. The fatigue damage of PUM exhibits an obvious three-stage damage law: the rapid development stage (accounting for about 10–20% of the fatigue life), the deformation stability expansion stage (accounting for about 70–80% of the fatigue life), and the instability development stage (accounting for about 10–20% of the fatigue life). The fatigue damage factors (D_B) were calculated based on stiffness, according to $D_B = \frac{EI_0 - EI_{nr}}{EI_0 - EI_{Nr}}$, and the fatigue damage functions of PUM were fitted based on the stiffness degradation, according to $f\left(\frac{n}{N}\right) = 1 - \frac{1-(n/N)^a}{(1-n/N)^b}$. The fatigue damage fitting curves have good correlation with the calculation results of the damage factor based on test data, which can predict the stiffness degradation of PUM at different load levels. The results can help further the understanding of the fatigue characteristics and damage mechanism of PUM, which will provide theoretical support for the application of PUM in pavement structures.

Keywords: polyurethane; mixture; fatigue; stiffness; four-point bending fatigue tests; damage factor

Citation: Sun, M.; Qu, G.; Geng, L.; Hou, D.; Jing, S. Fatigue Properties and Damage Characteristics of Polyurethane Mixtures under a Stress Control Mode. *Sustainability* **2022**, *14*, 10966. https://doi.org/10.3390/su141710966

Academic Editors: Dawei Wang, Chaohui Wang, Kai Liu and Qian Chen

Received: 11 August 2022
Accepted: 31 August 2022
Published: 2 September 2022

1. Introduction

Polyurethane is the general name for polymers containing repeated carbamate groups (-NHCOO-) in the main chain [1]. It is a "designable" polymer elastomer material with intermediate properties of plastic and rubber [2]. A polyurethane (PU) binder can solidify with hydroxyl groups in air and on the aggregate surface to form urea and a bonded network structure, which increases the strength of PUM continuously [3]. Thus, the temperature stability, mechanical properties, and fatigue properties of polyurethane mixtures essentially differ from those of asphalt mixtures. Due to its temperature stability, high elasticity, and wear resistance, many researchers have proposed the use of polyurethane (PU) instead of asphalt as a pavement binder [4,5]. Polyurethane is an environmentally friendly pavement binder; the production, transportation, and compaction of PU mixtures (PUMs) can be carried out at room temperature, which can greatly aid in global environmental conservation and ecologically sustainable development [6]. Since 2017, researchers have begun to use polyurethane binders to prepare polyurethane mixture to replace the mixture of open-graded friction course (OGFC), so as to solve the problems of the easy dispersion, poor fatigue stability, and poor permeability retention of OGFC mixture [4–6]. Then, more and more research on the application of polyurethane mixture in functional pavement was carried out [7–14]. In addition, in recent years, with the increasing demand for long-life pavement, some scholars adopted PU instead of asphalt as a binder to improve the service life and stability of pavement [11]. The PU mixture with an interlocking structure shows excellent temperature stability, fatigue stability, and mechanical properties [9]. Otherwise, polyurethane concrete for steel bridge deck pavement is developed according to the mechanical characteristics of steel bridge decks and the performance requirements of steel deck pavement material [15]. At present, PU mixture is mainly studied in three aspects: functional pavement, pavement structural layer, and bridge deck pavement. The composition design, curing reaction mechanism, performance evaluation, and construction technology of PUM are being studied by an increasing number of researchers worldwide [16–18].

The fatigue damage of pavement structures is most often caused by fatigue damage to its constituent materials [19]. To ensure the service life and safe performance of pavement structures, the fatigue characteristics of pavement materials must be tested and evaluated accurately [20]. Therefore, a variety of different fatigue tests have been carried out by many researchers to analyze the anti-fatigue properties of asphalt mixture and cement concrete [21]. Maalej et al. studied the behavior of hybrid-fiber engineered cementitious composites when subjected to dynamic tensile loading and impact loading [22]. The fatigue properties of engineered cementitious composites with the characteristic of low drying shrinkage have been investigated by Zhang et al. [23]. The plateau valve of the stiffness modulus degradation ratio curve was calculated to detect the fatigue damage characteristics of asphalt mixture by Guo et al. [24]. Shen et al. used the characteristic value of the damage curve as the evaluation index for rubber asphalt mixture fatigue performance, and established a prediction model for the BP neural network full-cycle fatigue life [25]. Huang et al. found that the fatigue self-healing efficiency is proportional to the asphalt and self-healing time, and inversely proportional to the extent of the void ratio, damage, and strain [26]. Menozzi et al. studied the fatigue damage and self-healing mechanism of different asphalt mixtures [27]. The aforementioned studies reveal the fatigue properties of asphalt mixtures and establish different fatigue models to calculate their properties under different fatigue loading modes. However, the research on the fatigue damage of PUM is relatively limited. Thus, the fatigue properties of PUM are still not understood adequately, along with the failure mode and damage mechanism, which is an obstacle to the popularization and application of PUM in pavement engineering.

Therefore, the fatigue properties and fatigue damage models of PUM need to be investigated in detail to improve engineering applications. The four-point bending static load tests, fatigue tests, and digital speckle correlation method (DSCM) tests at different control load levels of PUM and SMA were carried out in this paper. The failure strength, midspan deflection, fatigue life, and development law of the stiffness modulus, midspan deflection, and maximum tensile strain were detected and evaluated. The calculation method of the fatigue damage factor was established based on the stiffness modulus, and the fatigue damage factor development law of PUM was analyzed. Based on the composite material fatigue damage model, the fatigue damage prediction models of PUM based on stiffness degradation were established. The results provide a comprehensive understanding of the PUM fatigue characteristics and promote its application in pavement engineering.

2. Materials and Methods

2.1. Materials

The materials used mainly include a one-component wet curing PU binder, styrene butadiene styrene (SBS)-modified asphalt, and aggregates. The SBS-modified asphalt is often used at present, thus the performance of SBS-modified asphalt mixture was chosen for comparative analysis. The technical indexes of SBS-modified asphalt meet the requirements of I-D in JTG F40-2004 and can be observed in Table 1. The PU binder is a modified isocyanate containing a certain amount of terminal NCO groups (i.e., isocyanate group), which is prepared by Wanhua Chemical Co., Ltd. (Yantai, China). The indexes of the PU binder are shown in Table 2.

Table 1. The indexes of SBS-modified asphalt.

Technical Indicators	Test Method	Unit	Typical Value
Penetration	T 0604	0.1 mm	53
Penetration Index	T 0604	/	−0.5
5 °C Ductility	T 0605	cm	48
Softening Point	T 0606	°C	86.5
Dynamic Viscosity of 135 °C	T 0625	Pa.s	2.16

Table 2. The indexes of PU binder.

Technical Indicators	Test Method	Unit	Technical Requirement	Typical Value
Appearance	GB/T 13658	/	Brownish yellow liquid	/
Viscosity/25 °C	GB/T 12009.3	mPa.s	1200–2200	1600
Density/25 °C	GB/T 4472	g/cm^3	1.05~1.11	1.08
Acidity value	GB/T 12009.5	/	≤0.03	0.01
Surface dry time/25 °C, 50% Humidity	GB/T 13477.5	h	5–10	6
Tensile strength/25 °C	GB/T 16777	MPa	≥10	26
Elongation at break/25 °C	GB/T 16777	%	≥100	200

2.2. Mixture Composition Design

The aggregates of 0–3 mm, 5–10 mm, and 10–15 mm, along with mineral powder, both the PU mixtures (PUMs), and the stone mastic asphalt mixture (SMA) formed a skeleton dense structure; however, the PUM was designed according to the requirements of asphalt macadam concrete and the SMA was designed according to the requirements of stone mastic asphalt mixtures. The optimum dosage of PU and asphalt binder was determined by the Marshall mix design method. The composition design results of PUM with different aggregate nominal maximum sizes are shown in Table 3. The nominal maximum particle sizes of 13.2 mm and 9.5 mm are written as 13 and 10, respectively. The basic performance indexes of the three mixtures are shown in Table 4.

Table 3. Material composition design results.

Sieve Size (mm)	Cumulative Passing Percentage of Each Sieve (mm)/%										Binder Content/%
	16.0	13.2	9.5	4.75	2.36	1.18	0.6	0.3	0.15	0.075	
SMA-10	100.0	100.0	98.7	43.9	25.2	21.8	18.6	16.3	14.0	12.1	7.0
PUM-10	100.0	100.0	98.7	43.9	25.2	21.8	18.6	16.3	14.0	12.1	6.5
PUM-13	100.0	95.6	66.8	31.2	21.5	16.7	11.5	8.6	6.3	2.5	6.3

Table 4. The basic performance indexes of various mixtures.

Technical Indicators	Unit	SMA-10	PUM-10	PUM-13
Void ratio (VV)	%	5.3	4.5	4.8
Voids in the Mineral Aggregate (VMA)	%	17.4	14.5	14.8
Marshall Stability (MS)	KN	56.8	54.7	55.4
Dynamic stability/60 °C	Times/mm^{-1}	9146	82540	86327
Low temperature bending strain/−10 °C	με	3120	5734	5628
Splitting strength	MPa	1.38	3.05	3.21

3. Methodology

3.1. Experimental Scheme

3.1.1. Preparation of the Test Specimens and Preservation

The preparation of test specimens was conducted as follows:

(1) The PUM-10 and PUM-13 were prepared at room temperature, and the SMA-10 was mixed in accordance with the "Standard Test Methods of Bitumen and Bituminous Mixtures for Highway Engineering".
(2) The evenly mixed mixture was loaded into a 500 mm × 500 mm × 70 mm test mold, and the test specimens were formed using the wheel grinding method.
(3) After curing at 15 °C for 4 days, the mold was removed, and the core was cut from the specimen. The size of the specimens was 380 mm × 65 mm × 50 mm, which were maintained at room temperature for 1 day.
(4) White matte primer was sprayed on the surface of the test specimens, followed by a matte black speckle; the finished specimens are shown in Figure 1.

Figure 1. Test specimens.

3.1.2. Ultimate Bending Static Test

The Instron 8801 electro-hydraulic servo loading system was used for four-point bending loading in the ultimate bending static test. The net span of the test beam was 300 mm, and the length of the mid-span pure bend was 100 mm. The four-point bending loading layout is shown in Figures 2 and 3. The displacement control mode was used for loading, and the loading rate was 0.01 mm/s. During the loading process, the time, displacement, force, and bending stress were recorded.

3.1.3. Four-Point Bending Fatigue Test

In the stress control mode fatigue test, the obvious fatigue fracture of the specimens was taken as the failure state standard [23]. The stress control mode can describe a situation where the pavement stiffness is large, the surface strain increases rapidly, and the final pavement fatigue cracks precisely [24]. Therefore, in this paper, the four-point bending fatigue tests of mixtures under different control load levels were carried out to evaluate the fatigue characteristics.

Figure 2. Four-point bending loading layout.

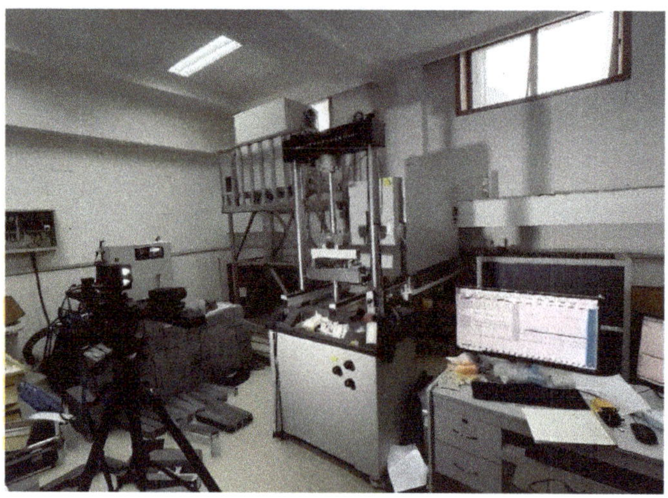

Figure 3. Fatigue loading and DSCM acquisition site.

The four-point bending loading method was adopted to obtain an approximate one-dimensional force in the middle portion (i.e., the pure bending section). The fatigue loading and DSCM collection site are shown in Figure 3. One end of the test support rolls and the other end slides; the spacing is 300 mm, the loading spacing is 100 mm, and the supports are arranged on the left and right 1/3rd of the span.

(1) Load waveform

Many studies have shown that the effect of the dynamic loading mode imposed on a road surface by a car driving on the road is similar to that of a sine wave [10]. When the frequency of the loading waveform is 10 Hz, it corresponds to the axle load effect of a vehicle traveling at 60–65 km/h [11]. Therefore, in this study, uninterrupted asymmetric equal-amplitude sine wave loading was used with a loading frequency of 10 Hz.

(2) Load level

The load level is the stress level ratio; many previous studies have shown that, in comparison to SMA, PUM is suitable for structural layers with higher load levels [5,6]. Therefore, the load levels of SMA-10 were determined to be 0.3, 0.4, and 0.5, and the load levels of PUM-10 and PUM-13 were determined to be 0.5, 0.6, and 0.7.

(3) Fatigue load cycle eigenvalues

The fatigue load cycle eigenvalue reflects the magnitude of the force change on the section [9]. The fatigue tests on relevant materials have shown that the fatigue loading cycle eigenvalue (ρ) is an important factor affecting fatigue failure [10]. The fatigue cycle eigenvalues of most pavement materials lie between 0.1 and 0.3, and the ρ value of the four-point bending fatigue is taken as 0.1. The specific load parameters of the four-point bending fatigue test are summarized in Table 5.

Table 5. Specific load parameters.

Test Beam Number	Fatigue Loading Data			Static Load Maximum Force/KN	Loading Frequency/ Hz
	Loading Level	Fatigue Lower Limit P_{min}/kN	Fatigue Ceiling P_{max}/kN		
SMA-10	0.3	0.06	0.60	2	10
	0.4	0.08	0.80	2	10
	0.5	0.10	1.00	2	10
PUM-10	0.5	0.41	4.05	8.1	10
	0.6	0.49	4.86	8.1	10
	0.7	0.57	5.67	8.1	10
PUM-13	0.5	0.45	4.45	8.9	10
	0.6	0.53	5.34	8.9	10
	0.7	0.62	6.23	8.9	10

(4) Loading process and method

First, preload the sample to the minimum load (0.1 kN) to ensure that the indenter is in close contact with the specimen and will not be emptied; further, observe the working status of each piece of equipment [24]. Second, apply the initial static load. Before the formal fatigue test, load the sample gradually to the median value of the fatigue limit. The static load takes 20 s to complete initial DSCM data acquisition. Then, start the sine wave fatigue loading, and perform static load tests after a certain number of loadings [25]. The static load value is the median value of the fatigue limit, which lasts for 20 s. During this process, complete the DSCM data collection. Repeat fatigue loading and dynamic loading until the specimen fails. The specific loading procedure is shown in Figure 4.

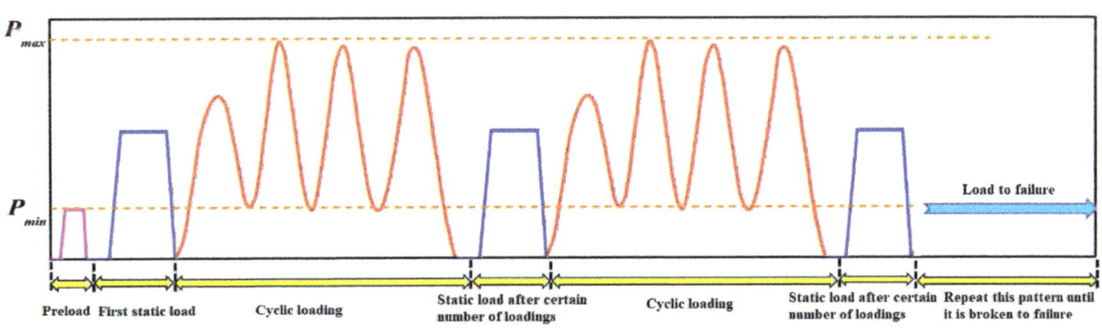

Figure 4. Fatigue test loading procedure. In the figure: The blue line indicates the static load applied, and the red line indicates the sinusoidal fluctuating load applied.

3.1.4. The DSCM Test

The DSCM equipment is a Xi'an Xintuo three-dimensional full-field strain measurement system. The equipment comprises an image acquisition and analysis system, two high-precision TAWOV cameras, a plane calibration board, a control box, a high-performance workstation, and a blue light source. The control box links the signals of the software and hardware. The TAWOV camera has a length of about 80 mm, a lens focal length of 25 mm, a resolution of 2448 × 2048, and 5 million pixels; the fixed included angle between the two cameras is 25°.

3.2. Calculation Method

3.2.1. Calculation Method of a Simply Supported Beam with Equal Moments

According to the force analysis of a simply supported beam with equal bending moments, the parameters—such as the loading, stiffness, and deflection—can be calculated. The destructive characteristics of PUM are similar to those of cement concrete, so the stiffness is used to evaluate the bearing capacity of PUM specimens [24,26]. The loading method of the test beam and its cross-sectional size are shown in Figure 5. It can be observed that this is a typical case of a simply supported beam with a constant bending moment, so the equations for the moment of inertia and deflection can be obtained, as shown in Equations (1) and (2), respectively.

$$I = \frac{1}{12}bh^3 \tag{1}$$

$$\omega_{max} = \frac{Fa}{24EI}\left(3l^2 - 4a^2\right) \tag{2}$$

where b is the width of the rectangular section, mm; h is the section height, mm; l is the simply supported beam span, mm; F is the load, N; a is the distance from the loading point to the fulcrum, mm; E is the elastic modulus, MPa; I is the moment of inertia to the main shaft, mm^4; ω_{max} is the deflection, mm; and $a = l/3$. The calculation formula for stiffness (EI) can then be obtained, as shown in Equation (3).

$$EI = \frac{Fa3l^2 - 4a^2}{24\omega_{max}} \tag{3}$$

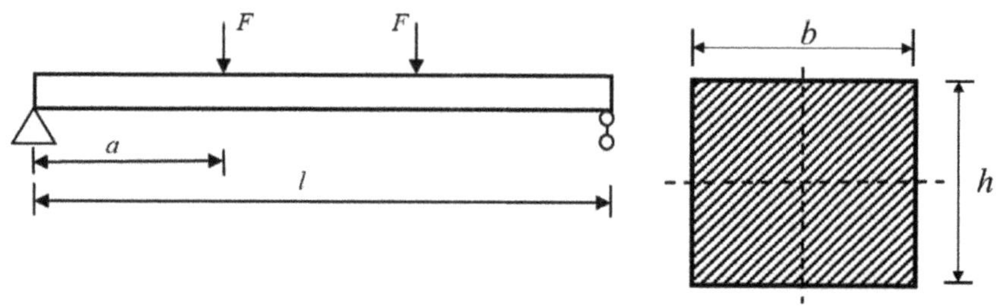

(a) (b)

Figure 5. Loading and cross-section schematics. (**a**) Loading diagram. (**b**) Loaded beam section.

3.2.2. Fatigue Damage Definition Based on Stiffness

Under repeated loading, the fatigue resistance of the beam decreases with increases in the number of loading cycles; therefore, the fatigue life of the beam continuously reduces, and the fatigue damage value is introduced to describe the damage status after different numbers of cycles [27]. The fatigue damage based on stiffness is defined in Equation (4).

$$D_B = \frac{EI_0 - EI_{nr}}{EI_0 - EI_{Nr}} \tag{4}$$

where D_B is the fatigue damage factor based on the residual stiffness, and its value range is [0, 1); EI_0 is the initial bending stiffness of the specimen; EI_{nr} is the bending stiffness of the specimen under the nth fatigue loading cycle; and EI_{Nr} is the bending stiffness of the specimen under the Nth fatigue loading cycle (that is, when the beam experiences fatigue failure). A functional relationship between the fatigue damage factor (D_B) and the fatigue life ratio *n/N* is established, as shown in Equation (5).

$$D_B = \frac{EI_0 - EI_{nr}}{EI_0 - EI_{Nr}} = f\left(\frac{n}{N}\right) \quad (5)$$

4. Results and Discussion

4.1. Static Load Failure Characteristics

The relationship of the span deflection with the bending force between the three mixing materials is shown in Figure 6. The bending strength test results of the three materials under static loading are shown in Table 6.

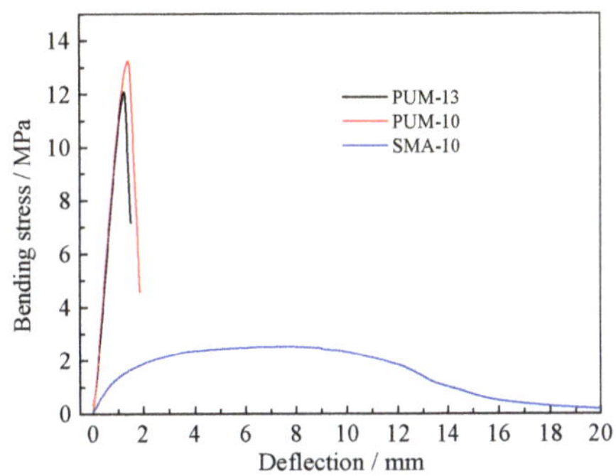

Figure 6. Midspan deflection and bending stress curves.

Table 6. Static load test results of three mixtures.

Type of Mixture	Static Load Failure Time/s	Maximum Deflection/mm	Bending Failure Stress/MPa
SMA-10	44.3	20.0	2.9
PUM-10	2.36	1.70	13.2
PUM-13	1.8	1.36	12.1

According to the deformation curves of the three mixtures, the bending failure processes of PUM and SMA differ under static loading. The bending failure of SMA-13 experiences an elastic stage, strain hardening stage, and stress relaxation stage, while the curves of PUM-10 and PUM-13 only experience an elastic stage and stress relaxation stage; the characteristics of the elastic stage are obvious, and the bending stress increases with increases in the mid-span deflection in this stage, which has an approximately linear trend.

The deformation ability of the PUM test specimens is very weak. The mid-span deflections of PUM-10 and PUM-13 are less than 2 mm, while the mid-span deflection of SMA-10 reaches 20 mm. However, the bending failure strength of PUM is significantly higher than that of SMA, and the bending failure strengths of PUM-10 and PUM-13 are about four times that of SMA-10. The bending load time of PUM-10 and PUM-13 is about 2 s, while that of SMA-10 is about 44 s. This indicates that the failure mode of PUM under bending static loads is brittle failure, which is similar to that of cement concrete [22,23]. The reason is analyzed that the urea bond is formed by the reaction of isocyanate groups (NCO) in the PU binder and water, then the urea bond and metal oxide chelate to form the urea metal oxide complex due to the hydrogen bond, so that the PUMs have high strength and strong load resistance. However, when the failure load of the PUMs is reached,

As shown in Figure 10, the fatigue failure law of PUM presents an obvious S-shaped three-stage development curve. The first stage is the development stage, where the stiffness is greatly reduced within the first 10–20% of the fatigue life; the second stage is the stable stage, which accounts for about 70–80% of the entire fatigue life, where the stiffness of the beam body maintains a relatively gentle attenuation; the third stage experiences the sudden changes in failure and accounts for about 10–20% of the entire fatigue process, where the structure is characterized by brittle fatigue fracture [26].

The load level has a significant effect on the three-stage history of the fatigue failure of PUM. At higher load levels, the development stage of the fatigue curve has a longer duration, the rate of reduction in stiffness is faster in the stable stage, and the reduction in stiffness in the brittle fatigue fracture stage is larger [27]. It is intuitively known that the slopes of the stable stage and brittle fatigue fracture stage of the stiffness curve at higher load levels are larger; that is, the slope of the stiffness curve at a 0.7 load level > 0.6 load level > 0.5 load level.

The stable stage of SMA takes a long time to complete, and the distinction between the failure stage and the stable stage is not obvious. This is because the fatigue failure form of SMA is viscoelastic damage at the tested temperature [11].

4.2.3. Development Law of the Mid-Span Deflection

The deflection change reflects the overall fatigue performance of the structure. Figure 11 shows the curve between the loading cycle ratio and deflection of PUM-10 and PUM-13. It can be observed from the figure that the deflection curve exhibits the following three stages: the rapid development stage, the deformation stability expansion stage, and the instability development stage; that is, the mid-span deflection develops obviously in the first 10% of cyclic loading, the mid-span deflection increases slowly during 20–80% of cyclic loading, and within 90–100% of cyclic loading, the fatigue failure stage is completed; the deflection subsequently enters the rapid development stage once again [12,13]. At lower load levels, as shown in the deflection curve of the 0.5 load level, the deflection of the test beam changes greatly at the initial stages of fatigue loading and at the time of fatigue fracture; that is, the front and rear sections of the curve are relatively short and the middle section is long. However, at a higher load level, as shown in the deflection curve of the 0.7 load level, the rapid fatigue development stage and fatigue failure stage last a relatively long time.

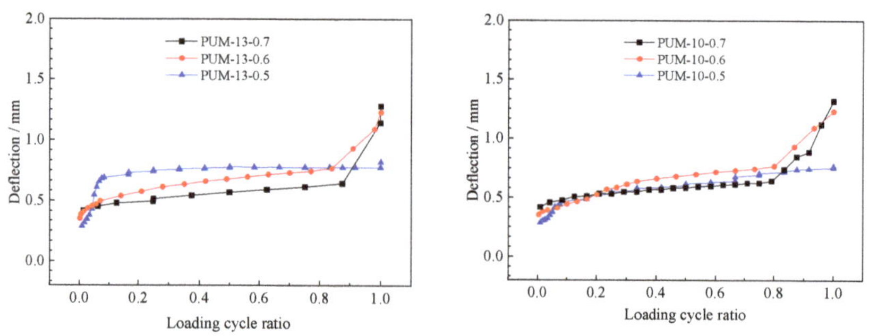

(**a**) (**b**)

Figure 11. Deflection curve at different loading cycle ratios. (**a**) Deflection curve of PUM-13. (**b**) Deflection curve of PUM-10.

4.3. Analysis of the Maximum Tensile Strain Based on DSCM Data Acquisition

Strain is also an evaluation index for the deformation capacity of asphalt materials. A series of strain analyses were performed on the asphalt mixture by using DSCM equipment. Figure 12 shows the curve between the maximum tensile strain at the bottom and the loading cycle ratio.

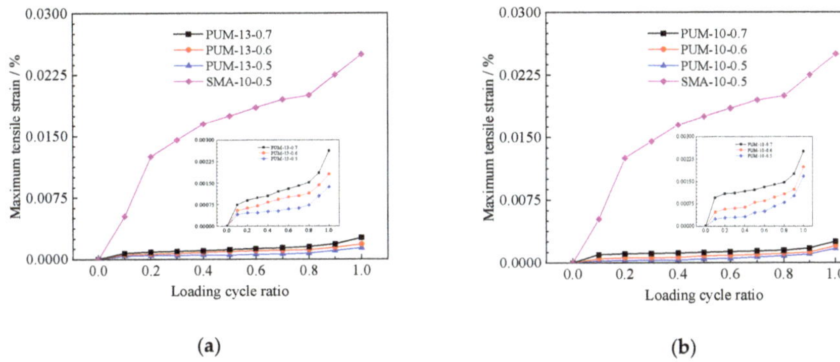

Figure 12. Maximum tensile strain curve with loading cycle ratios. (**a**) Maximum tensile strain curve of PUM-13. (**b**) Maximum tensile strain curve of PUM-10.

It can be observed from Figure 12 that the maximum tensile strain and development speed of SMA-10 are much higher than those of PUM-10 and PUM-13. The development of the maximum tensile strain of PUM and SMA exhibited a development trend of "three stages": the rapid development stage, the deformation stability expansion stage, and the instability development stage. It can be observed that the PUM specimens undergo rapid bending deformation under loading, cracks and expansion begin to occur at the internal defect positions, and the rate of change subsequently slows down. This is because the PU binder acts as a bridge to bear part of the load and prevents the specimen from continuing to crack [7]. However, when reaching the third stage, the tensile strain increases sharply from approximately horizontal development to failure, and the fracture mode is similar to brittle fracture. The failure modes of PUM at different loading levels are similar. This type of failure mode has a great impact on the structural safety, and it is necessary to carry out research on the relevant damage mechanism to accurately predict the fatigue characteristics.

4.4. Fatigue Damage Characteristics of PUM

4.4.1. Development Law of Fatigue Damage Based on Stiffness

The fatigue damage factor of PUM under different loading levels was calculated according to Equation (5). Figure 13 shows the fatigue damage factor curves under different loading cycle ratios.

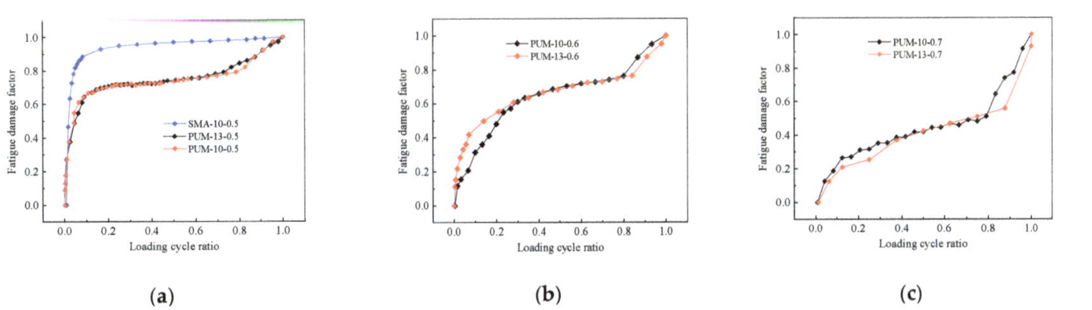

Figure 13. Fatigue damage factor curves under different loading cycle ratios. (**a**) Loading level of 0.5. (**b**) Loading level of 0.6. (**c**) Loading level of 0.7.

It can be observed from Figure 13 that the damage curves of PUM and SMA are different. At the loading level of 0.5, SMA-10 exhibited two-stage damage characteristics, and the 0.9 damage factor was completed at a loading cycle ratio of about 20%; the stage of slow development of fatigue damage until failure was then entered. The reason for this phenomenon is that the load level of 0.5 is a high stress level for SMA, which leads to a large amount of damage in the early stages of loading [6,28].

The fatigue damage factor curves with the three loading levels of PUM-10 and PUM-13 exhibit the same pattern; that is, there is a rapid increase in the initial stages of life, which is caused by the initial damage to the material and the formation of matrix cracks at the beginning of the fatigue life. When the matrix cracks reach a certain density, the characteristic damage state appears, the matrix cracks increase no more, and the damage factor growth becomes moderate, basically exhibiting a linear increase. In the final stages of life, damage accumulates rapidly and evolves intensely. Furthermore, the higher the load level of PUM, the faster the development of the concentrated evolution stage of the damage factor curve.

It may be that SBS-modified asphalt is a kind of viscoelastic binder, so SMA-10 has strong deformation capacity under load. When the fatigue damage factor reaches 0.9, it can still bear the load. The PU binder bonds the aggregate through the curing reaction, thus the PUM has high strength and relatively low deformation resistance. The fatigue damage factor develops relatively slowly under cyclic load, however, when the failure load of the PUMs is reached, the interface between the PU binder and aggregates, or the aggregate in the mixture, enters the final stage of life, and damage accumulates rapidly and evolves intensely.

4.4.2. Fatigue Damage Function Construction Based on Stiffness Degradation

Determining the fatigue damage function $f\left(\frac{n}{N}\right)$ based on stiffness degradation is the key to analyzing the fatigue damage of mixtures; the fatigue test on PUM in this study also exhibited a typical three-stage stiffness degradation law [29,30], as shown in Figures 9 and 10. Therefore, the function must be able to clearly describe the three stages of the fatigue failure process; that is, the stiffness in the early stages of fatigue loading exhibits a relatively large decrease and then enters a linear stage of stable development. Subsequently, the stiffness exhibits a relatively large decrease before and until structural failure [30]. Many studies have shown that the functional form of Equation (6) has achieved good results in terms of fitting the fatigue damage of composite materials; therefore, this study used this model to fit the fatigue damage function of PUM [29].

$$f\left(\frac{n}{N}\right) = 1 - \frac{1-(n/N)^a}{(1-n/N)^b} \quad (6)$$

where a and b are the parameters to be fitted.

From the analysis of the stiffness degradation results, it is clear that the fatigue damage factors of PUM-10 and PUM-13 are similar at the same load levels, indicating that the maximum engineering particle size has little effect on the damage factors of PUM [30,31]. Therefore, the fatigue damage of the two graded PUM under the same loading level is described by the same function, and the damage factor under the same loading cycle is the average value of PUM-10 and PUM-13.

The fatigue damage of PUM under different load levels in the cyclic loading process is fitted according to Equation (6), and the fitting parameters are shown in Table 6. Further, the residual stiffness ratios at fatigue failure under different loading levels are summarized in Table 7; the fatigue damage function after fitting and the test results are plotted in Figure 14.

The fitting curves are in good agreement with the experimental data, indicating that the fitting formulas can be used to predict the PUM fatigue damage at different loading levels [32]. The ratio of the residual stiffness at fatigue failure at different loading levels shows that the higher the load level, the greater the residual stiffness at fatigue failure [33].

Table 7. Degradation function and parameters fitted of PUM stiffness.

Loading Level	a	b	Fatigue Damage Function	Correlation Coefficient	EI_{Nr}/EI_0
0.5	0.19	0.79	$f\left(\frac{n}{N}\right) = 1 - \frac{1-(n/N)^{0.19}}{(1-n/N)^{0.79}}$	0.9967	0.198
0.6	0.31	0.67	$f\left(\frac{n}{N}\right) = 1 - \frac{1-(n/N)^{0.31}}{(1-n/N)^{0.67}}$	0.9832	0.287
0.7	0.59	0.76	$f\left(\frac{n}{N}\right) = 1 - \frac{1-(n/N)^{0.59}}{(1-n/N)^{0.76}}$	0.9878	0.403

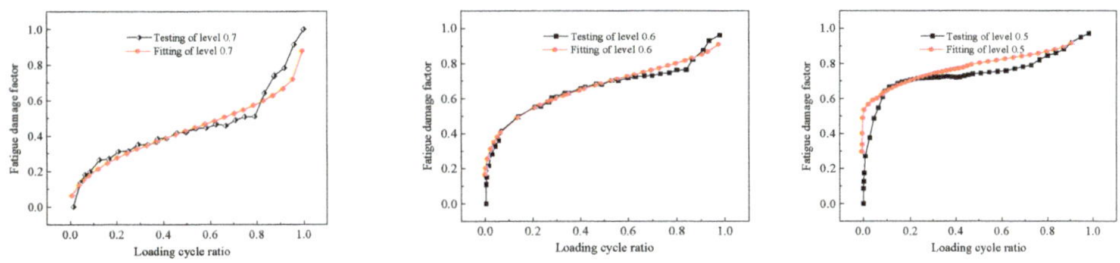

(a) (b) (c)

Figure 14. Fatigue damage factor curves at different loading cycle rates. (**a**) Loading level of 0.5. (**b**) Loading level of 0.6. (**c**) Loading level of 0.7.

5. Conclusions

(1) The bending failure processes of PUM and SMA differ under static loading, and the curves of PUM only go through an elastic stage and stress relaxation stage. The static failure mode and fatigue damage of PUM is brittle failure.

(2) The fatigue life of PUM is much larger than that of SMA under the same bending stress. The fatigue damage of PUM exhibits an obvious three-stage damage law and the fatigue damage leads to brittle failure.

(3) The deflection development and maximum strain development increment of PUM increase significantly at the beginning of the fatigue cycle. With increases in the fatigue load cycles, the growth rate slows down and enters a relatively stable development stage, and increases greatly near failure.

(4) The calculation method of the fatigue damage factor based on stiffness degradation is constructed, and the fatigue damage functions of PUM at different load levels are fitted, which can help to predict the stiffness degradation of PUM.

(5) In this paper, four-point bending tests with three load levels are carried out. It is necessary to carry out four-point bending fatigue tests with more load levels, so as to fit the S-N curve of PUM according to the fatigue life and applied stress amplitude. This can help predict the fatigue life of PUM.

Author Contributions: Conceptualization, M.S. and G.Q.; methodology, S.J.; software, L.G.; validation, D.H. and S.J.; formal analysis, G.Q.; investigation, S.J.; resources, D.H.; data curation, L.G.; writing—original draft preparation, M.S.; writing—review and editing, G.Q.; visualization, L.G.; supervision, M.S.; project administration, M.S.; funding acquisition, M.S. All authors have read and agreed to the published version of the manuscript.

Funding: This work was supported by Key Scientific Research Projects in the Transportation Industry of the Ministry of Transport (2019MS2028), Shandong Expressway Group Project (HSB 2021-72), Open Project of Shandong Key Laboratory of Highway Technology and Safety Assessment (SH202107), Key Program of Natural Science Foundation of Shandong Province (ZR2020KE007), and Shandong Provincial Young Scholars Innovative Research Team Development Program in Colleges and Universities (2019KJG004).

Conflicts of Interest: The authors declare no conflict of interest.

References

1. Chem, Y.C.; Tseng, S.M.; Hsieh, K.H. Damping properties of interpenetrating polymer networks of polyurethane-modified epoxy and polyurethanes. *J. Appl. Polym. Sci.* **2015**, *74*, 328–335.
2. Wang, Q.; Chen, S.; Wang, T.; Zhang, X. Damping, thermal, and mechanical properties of polyurethane based on poly (tetramethylene glycol)/epoxy interpenetrating polymer networks: Effects of composition and isocyanate index. *Appl. Phys. A* **2011**, *104*, 375–382. [CrossRef]
3. Wang, J.; Yuan, J.; Xiao, F.; Li, Z.; Wang, J.; Xu, Z. Performance investigation and sustainability evaluation of multiple-polymer asphalt mixtures in airfield pavement. *J. Clean. Prod.* **2018**, *189*, 67–77.
4. Chen, J.; Yin, X.; Wang, H.; Ding, Y. Evaluation of durability functional performance of porous polyurethane mixture in porous pavement. *J. Clean. Prod.* **2018**, *188*, 12–19.
5. Chen, J.; Ma, X.; Wang, H.; Xie, P.; Huang, W. Experimental study on anti-icing and deicing performance of polyurethane concrete as road surface layer. *Constr. Build. Mater.* **2018**, *161*, 598–605. [CrossRef]
6. Chen, J.; Yao, C.; Wang, H.; Huang, W.; Ma, X.; Qian, J. Interface Shear Performance between Porous Polyurethane Mixture and Asphalt Sublayer. *Appl. Sci.* **2018**, *8*, 623. [CrossRef]
7. Wang, D.; Schacht, A.; Leng, Z.; Leng, C.; Kollmann, J.; Oeser, M. Effects of Materials Composition on Mechanical and Acoustic Performance of Poroelastic Road Surface (PERS). *Constr. Build. Mater.* **2017**, *135*, 352–360. [CrossRef]
8. Wang, D.; Liu, P.; Leng, Z.; Leng, C.; Lu, G.; Buch, M.; Oeser, M. Suitability of Poroelatic Road Surface (PERS) for Urban Roads in Cold Regions: Mechanical Functional Performance Assessment. *J. Clean. Prod.* **2017**, *165*, 1340–1350. [CrossRef]
9. Sun, M.; Bi, Y.; Zheng, M.; Wang, J.; Wang, L. Performance of polyurethane mixtures with skeleton-interlocking structure. *J. Mater. Civ. Eng.* **2020**, *32*, 04019358. [CrossRef]
10. Lu, G.; Renken, L.; Li, T.; Wang, D.; Li, H.; Oeser, M. Experimental study on the polyurethane-bound pervious mixtures in the application of permeable pavement. *Constr. Build. Mater.* **2019**, *202*, 838–850. [CrossRef]
11. Li, T.; Lu, G.; Wang, D.; Hong, B.; Tan, Y.; Oeser, M. Key Properties of High-performance Polyurethane Bounded Pervious Mixture. *China J. Highw. Transp.* **2019**, *32*, 158–169. (In Chinese)
12. Lu, G.; Liu, P.; Wang, Y.; Faßbender, S.; Wang, D.; Oeser, M. Development of a sustainable pervious pavement material using recycled ceramic aggregate bio-based polyurethane binder. *J. Clean. Prod.* **2019**, *220*, 1052–1060. [CrossRef]
13. Törzs, T.; Lu, G.; Monteiro, A.O.; Wang, D.; Grabe, J.; Oeser, M. Hydraulic properties of polyurethane-bound permeable pavement materials considering unsaturated flow. *Constr. Build. Mater.* **2019**, *212*, 422–430. [CrossRef]
14. Lu, G.; Liu, P.; Törzs, T.; Wang, D.; Oeser, M.; Grabe, J. Numerical analysis for the influence of saturation on the base course of permeable pavement with a novel polyurethane binder. *Constr. Build. Mater.* **2020**, *240*, 1–9. [CrossRef]
15. Xu, S.; Zhang, Y.; Guo, Y.; Ma, C.; Gao, D.; Peng, G. Determination of polyurethane concrete compaction timing based on penetration resistance test system. *China J. Highw. Transp.* **2021**, *34*, 226–235. (In Chinese)
16. Sun, M.; Bi, Y.; Zhuang, W.; Chen, S.; Zhao, P.; Pang, D.; Zhang, W. Mechanism of polyurethane binder curing reaction and evaluation of polyurethane mixture properties. *Coatings* **2021**, *11*, 1454. [CrossRef]
17. Cong, L.; Wang, T.; Tan, L.; Yuan, J.; Shi, J. Laboratory evaluation on performance of porous polyurethane mixtures and OGFC. *Constr. Build. Mater.* **2018**, *169*, 436–442. [CrossRef]
18. Izaks, R.; Rathore, M.; Haritonovs, V.; Zaumanis, M. Performance properties of high modulus asphalt concrete containing high reclaimed asphalt content and polymer modified binder. *Int. J. Pavement Eng.* **2022**, *23*, 2255–2264. [CrossRef]
19. Yoshida, N. Fatigue failure criterion for hydraulic graded iron and steel slag base-course and ITS implementation into the Japanese asphalt pavement design method. *Struct. Eng.* **2014**, *2*, 62–68. [CrossRef]
20. Lv, S.; Luo, Z.; Xie, J. Fatigue performance of aging asphalt mixtures. *Polimery* **2015**, *60*, 126–131. [CrossRef]
21. Lv, S.; Liu, C.; Yao, H.; Zheng, J. Comparisons of synchronous measurement methods on various moduli of asphalt mixtures. *Constr. Build. Mater.* **2018**, *158*, 1035–1045. [CrossRef]
22. Maalej, M.; Quek, S.T.; Zhang, J. Behavior of Hybrid-Fiber Engineered Cementitious Composites Subjected to Dynamic Tensile Loading and Project Impact. *J. Mater. Civ. Eng. ASCE* **2005**, *17*, 143–152. [CrossRef]
23. Zhang, J.; Gong, C.; Guo, Z.; Zhang, M. Engineered Cementitious Composites with Characteristic of Low Drying Shrinkage. *Cem. Concr. Res.* **2009**, *39*, 303–312. [CrossRef]
24. Guo, Z.; Wang, L.; Feng, L.; Guo, Y. Research on fatigue performance of composite crumb rubber modified asphalt mixture under freeze thaw cycles. *Constr. Build. Mater.* **2022**, *323*, 126603.
25. Shen, A.; Yu, M.; Zhou, X.; Lv, Z.; Song, P. Fatigue Damage and Full-cycle life estimation of rubber asphalt mixture. *J. Build. Mater. Struct.* **2018**, *21*, 620–625. [CrossRef]
26. Huang, W.; Peng, L.; Mao, Z.; Sun, Z. Analysis of foacors influencing fatigue self-healing performance of SBS modified asphalt mixture. *J. Build. Mater.* **2016**, *19*, 950–956. [CrossRef]
27. Menozzi, A.; Garcia, A.; Partl, M.N.; Tebaldi, G.; Schuetz, P. Induction healing of fatigue damage in asphalt test samples. *Constr. Build. Mater.* **2015**, *74*, 162–168. [CrossRef]
28. Zhu, H.; Yang, W. A survey on stiffness reduction models of fiber reinforced plastics under cyclic loading. *Adv. Mech.* **2002**, *32*, 69–80. (In Chinese)
29. Feng, P.; Du, S.; Wang, D. Fatigue residual stiffness degradation model for composite laminates. *Acta Mech. Solida Sin.* **2003**, *24*, 446–452. (In Chinese)

30. Zhu, H.; Yu, Z.; Sun, J. Experimental research on fatigue behavior of reinforced concrete T-beam. *J. Highw. Transp. Res. Dev.* **2013**, *30*, 53–58. (In Chinese)
31. Zhu, H.; Yao, Z.; Xiu, L.; Yu, Z. Reinforced Concrete Beam's stiffness degeneration regulation and its calculation formula under the action of fatigue load. *J. Civ. Archit. Environ. Eng.* **2014**, *36*, 1–5.
32. Cheng, L. Flexural fatigue analysis of a CFRP form reinforced concrete bridge deck. *Compos. Struct.* **2011**, *93*, 2895–2902. (In Chinese) [CrossRef]
33. Nie, J.; Wang, Y.; Cai, C. Experimental research on fatigue behavior of RC beams strengthened with steel plate-concrete composite technique. *J. Struct. Eng.* **2011**, *137*, 772–781. (In Chinese) [CrossRef]

sustainability

Article

Thermal Conductivity Evaluation and Road Performance Test of Steel Slag Asphalt Mixture

Yangsen Cao [1], Aimin Sha [2,*], Zhuangzhuang Liu [2,*], Fan Zhang [1], Jiarong Li [1] and Hai Liu [3]

1 School of Highway, Chang'an University, Xi'an 710064, China; yscao@chd.edu.cn (Y.C.); fanzhang@chd.edu.cn (F.Z.); jrl@chd.edu.cn (J.L.)
2 Key Laboratory for Special Area Highway Engineering of Ministry of Education, Chang'an University, South 2nd Ring Road Middle Section, Xi'an 710064, China
3 China Railway Siyuan Survey and Design Group Co., Ltd., Wuhan 430063, China; 2019021007@chd.edu.cn
* Correspondence: ams@chd.edu.cn (A.S.); zzliu@chd.edu (Z.L.)

Abstract: Substituting steel slag for mineral materials in road construction has potential economic and environmental benefits. Due to the excellent thermal conductivity of steel slag, it is often used in functional pavements. However, there are few studies on the thermal conductivity characterization of steel slag asphalt mixture (SSAM). For this reason, the thermal conductivity of SSAM was first qualitatively evaluated by microscopic characterizations. The thermal conductivity was the quantitatively evaluated by the heating wire method. Theoretical calculations were used to verify the reliability of the quantitative characterization. Finally, the effects of steel slag on the volume indices and the road performance of SSAM were studied. Results showed that active minerals such as iron oxides make the steel slag thermally conductive, while a large number of protrusions and micropores on the surface of the steel slag may be detrimental to thermal conductivity. The thermal conductivity first increases and then decreases with the steel slag content. The asphalt mixture with 60% steel slag replacing aggregate of 3–5 mm (6.6% of the mixture) had the highest thermal coefficient of 1.746 W/(m·°C), which is only 4.78% different from the theoretical value. The porosity and water absorption of SSAM gradually increased with the content of steel slag. The road performance test indicated that steel slag increased the high-temperature performance of the asphalt mixture to a certain extent, but weakened the low-temperature performance and moisture resistance. After comprehensive consideration of the thermal conductivity and road performance, it is recommended that the optimum content of steel slag is not more than 60%.

Keywords: steel slag asphalt mixture; microscopic characterization; thermal conductivity; road performance

Citation: Cao, Y.; Sha, A.; Liu, Z.; Zhang, F.; Li, J.; Liu, H. Thermal Conductivity Evaluation and Road Performance Test of Steel Slag Asphalt Mixture. *Sustainability* **2022**, *14*, 7288. https://doi.org/10.3390/su14127288

Academic Editors: Dawei Wang, Chaohui Wang, Kai Liu and Qian Chen

Received: 15 April 2022
Accepted: 11 June 2022
Published: 14 June 2022

1. Introduction

Clear waters and lush mountains are invaluable assets. Reducing the exploitation of natural resources is of great significance to global environmental protection. Transportation is a resource-intensive industry, in which the construction of transportation infrastructure consumes a lot of mineral resources. Seeking effective substitutes for ore resources has great potential for reducing mine development and promoting the sustainable development of society [1–4]. Steel slag is a by-product of the steelmaking process. It is a lumpy substance formed by the combination of residual flux and metal oxide in the steelmaking furnace after cooling, iron removal, and crushing. About 0.15 t of steel slag is produced for every 1 t of steel produced [5]. China is the world's largest steel producer, with annual steel production accounting for about 50% of global steel production capacity. In the past two years, the average annual discharge of steel slag is 100 million tons. The major industrial countries in the world have adopted the comprehensive utilization of steel slag as a strategy for green development [6]. Steel slag has the advantages of high strength, wear resistance, slip resistance, high basicity, and low cost, and has the potential for large-scale utilization in pavements [7,8].

Some research results have been obtained on the application of steel slag in road engineering, mainly involving two aspects: conventional performance and functional performance. Conventional properties are stiffness indicators of SSAMs, such as fatigue properties, expansion properties, aging properties, and other mechanically related properties. For example, Kavussi [9] studied the fatigue behavior of asphalt mixtures with electric arc furnace (EAF) steel slag as coarse aggregate in aged and unaged states. The four-point bending fatigue test showed that although the addition of EAF steel slag into the asphalt mixture improved the fatigue life of the sample, it had no obvious effect on the asphalt mixture in the aging state. To further improve the fatigue performance of the SSAM, Cao [10] modified the surface of the steel slag with cement paste and studied the fatigue life of the SSAM under soaking conditions. The results showed that the fatigue life of the modified SSAM was 23% higher than that of the original SSAM. The dense layer formed on the surface of the steel slag by the cement hydration products prevents the water from invading the steel slag, thereby improving the fatigue performance. Although the physical properties of steel slag meet the application requirements of asphalt mixtures, due to its angular shape, asphalt mixtures using steel slag as aggregates are prone to expansion and are affected by void problems. Mineral aggregates cannot be completely replaced with steel slag [11]. To reduce the volume expansion characteristics of steel slag and make full use of cold abandoned steel slag, Zhang [12] used limestone fine aggregate instead of steel slag fine aggregate to reduce the water swelling of cold abandoned SSAM. Masoudi [13,14] studied the long-term aging performance of the EAF steel slag warm mix asphalt mixture. Steel slag warm mix asphalt has better mechanical properties than limestone hot mix asphalt. The use of steel slag increases the aging index of the asphalt mixture and reduces the anti-aging ability of the asphalt mixture. In addition, the replacement of limestone aggregate with steel slag increases the optimal asphalt aggregate ratio by 0.45%. Zeng [15] studied the nonlinear viscoelastic–plastic deformation characteristics of SSAMs and proposed a new integral viscoelastic–plastic constitutive model combining the Schapery model and the improved Schwartz model. The model can more accurately reflect the elastic, viscoelastic, and viscoplastic deformation of SSAM in the creep process. Ameri et al. [16] studied the performance of steel slag in hot mix asphalt and warm mix asphalt. They showed that the hot mix asphalt mixture with steel slag instead of fine aggregate had the lowest elastic modulus and indirect tensile strength. However, using steel slag as coarse aggregate in warm mix asphalt improved the elastic modulus and indirect tensile strength of the mixture. Steel slag coarse aggregate is recommended for warm mix asphalt. To make full use of the coarse and fine aggregates of slag, Chen [17] discussed the comprehensive utilization technology of steel slag and iron slag, and used steel slag coarse aggregate, iron slag fine aggregate, and steel slag powder to replace 100% limestone. Some performance indicators of the asphalt mixture prepared by combining high-viscosity asphalt and the secondary mixing process are significantly improved. The high-temperature flow times and low-temperature fracture energy were increased by 18% and 23%, respectively. There are many methods for evaluating the rutting resistance of asphalt mixtures, such as rutting test, dynamic modulus test, creep test, etc. [18]. In addition, the rotary compaction process can also reflect the rutting resistance, more compaction times, and strong anti-rutting performance [19]. Huang [20] studied the compaction characteristics of steel slag porous asphalt mixture, and the results of rotary compaction showed that the number of compactions of porous asphalt mixture without steel slag was higher than that of steel slag porous asphalt mixture. It was indicated that the porous asphalt mixture without steel slag has a higher anti-rutting performance. However, Martinho's [21] research on the high-temperature performance of warm mix SSAM formed the opposite conclusion. In general, steel slag can improve the high-temperature performance of asphalt mixture. Liu [22] evaluated the feasibility of the porous asphalt mixture containing steel slag in the seasonal freezing area, and the results showed that the addition of steel slag can improve the pavement performance of the porous asphalt mixture. When the coarse aggregates are steel slag, the overall performance of the porous asphalt mixture is optimal. Crisman [23] compared the mechanical properties of rubber–asphalt mixture containing steel slag and ordinary asphalt mixture and found that the elastic modulus of steel slag rubber–asphalt mixture was higher than that of ordinary asphalt mixture at high temperature, but lower at low temperature. The permanent deformation resistance of steel slag rubber–asphalt mixture has been improved.

In terms of the functional utilization of steel slag, Li [24] studied the effect of steel slag on the electrical conductivity of graphite–carbon fiber composite conductive asphalt concrete and found that 100% steel slag content increased the conductivity of the mixture to 0.125 S/m. The contribution of the three materials to the conductivity followed the order carbon fiber, graphite, and steel slag (from highest to lowest). Xiang [25] studied the self-healing performance of SSAM under microwave heating conditions. The addition of steel slag can improve the healing effect of asphalt mixtures. The asphalt mixture of coarse stone and fine steel slag has a more uniform temperature distribution, and the healing effect is 1.11 times higher than that of asphalt mixtures without steel slag. Gao [26] mixed steel slag into asphalt mixture for microwave deicing, measured the temperature distribution and ice melting effect of SSAM, and verified the feasibility of ice melting of SSAM.

To sum up, the application of steel slag in road engineering mainly focuses on the basic properties of SSAM, such as fatigue, expansion, aging, and mechanical properties. This is because the primary task of SSAM is to meet the requirements of engineering construction. There is still insufficient research on the functional aspects of SSAM. Whether it is electrical conductivity, self-healing performance, or ice-melting performance, these properties are all related, to some extent, to the thermal conductivity of SSAM. Existing research rarely considers the basic indicators of its thermal conductivity. Further research into the thermal conductivity of SSAM is required.

Consequently, the goal of this paper is to study the thermal conductivity of SSAM, as well as the road performance of SSAMs with different thermal conductivity. It is expected that SSAM can meet the requirements of road performance while exerting thermal conductivity. Therefore, the thermal conductivity of steel slag was first qualitatively evaluated from a microscopic perspective. Second, the SSAM was prepared according to the principle of equal volume replacement and the heating test was conducted. The thermal conductivity of SSAM was quantitatively evaluated with the thermal coefficient as an index. Afterward, to prove the accuracy of the measured data, the thermal coefficient was estimated by the theoretical calculation method. Finally, the volume indices of the asphalt mixtures with different thermal conductivity were tested, and the road performances of the SSAM was verified.

2. Materials and Methods

The raw materials used to prepare the thermally conductive asphalt mixture mainly include asphalt, fillers, aggregates, and thermally conductive phase steel slag. The technical indicators of the raw materials are tested following the "Standard Test Methods of Bitumen and Bituminous Mixtures for Highway Engineering" (JTG E20-2011) [27]. The index standard comes from "Technical Specifications for Construction of Highway Asphalt Pavements" (JTG F40-2004) [28].

2.1. Materials

2.1.1. Asphalt

The test asphalt is SBS I-C modified asphalt; its technical indicators are shown in Table 1.

Table 1. Technical indicators of SBS I-C modified asphalt.

Technical Indicators		Unit	Standard	Test Results	Test Method
Penetration (25 °C, 100 g, 5 s)		0.1 mm	60–80	64.7	T0604-2011
Penetration index		-	≥−0.4	0.4	T0604-2011
Ductility (5 °C, cm/min)		cm	≥30	37.8	T0605-2011
Softening point (Ring-and-ball method)		°C	≥55	63.5	T0606-2011
Density		g/cm	-	1.014	T0603-2011
Dynamic viscosity (135 °C)		Pa·s	≤3	1.870	T0619-2011
Flash point (Cleveland Open Cup)		°C	≥230	260	T0611-2011
Solubility (Trichloroethylene)		%	≥99	99.7	T0607-2011
Elastic recovery (25 °C, 10 cm)		%	≥65	80	T0662-2000
Properties after rolling thin film oven test	Quality change	%	≤±1.0	−0.19	T0609-2011
	Penetration ratio (25 °C)	%	≥60	82.9	T0604-2011
	Ductility (5 °C)	cm	≥20	26.3	T0605-2011
	Softening Point (Ring-and-ball method)	°C	≥55	71.8	T0606-2011

2.1.2. Aggregates and Filler

The aggregates used in tests are mainly limestone, with a total of 4 fractions of particle sizes. The filler is alkaline ground limestone. The technical indicators of aggregates and filler are shown in Tables 2 and 3, respectively.

Table 2. Aggregate technical indicators.

Technical Indicators	Particle Size/mm				Standard	Test Method
	10–15	5–10	3–5	0–3		
Apparent relative density (g/cm^3)	2.716	2.727	2.745	2.695	≥2.5	T0304-2005
Gross volume relative density (g/cm^3)	2.684	2.685	2.658	2.647	-	T0304-2005
Crush value (%)	22.6	-	-	-	≤26	T0316-2005
Los Angeles abrasion value (%)	15.4	18.5	24.8	-	≤28	T0317-2005
Water absorption (%)	1.2	0.6	0.4	0.4	≤2	T0304-2005
Needle flake content (%)	10.69	8.01	-	-	≤12	T0312-2005

Table 3. Performance index of filler.

Technical Indicators	Test Results	Standard	Test Method
Apparent relative density (g/cm^3)	≥2.5	2.716	T0352-2000
Hydrophilic coefficient	<1	0.76	T0353-2000
Stability	No discoloration when heated	No color change	T0355-2000

2.1.3. Steel Slag

The steel slags in the research are basic oxygen furnace (BOF) steel slags. The steel slag replaces the aggregate in the asphalt mixture, which can form a coherent heat conduction path in the mixture, speed up the heat transfer rate in the mixture, and improve the thermal conductivity of the mixture. The performance indicators are shown in Table 4. The relative density of the steel slag of 0–3 mm in Table 4 is low, which may be because the fine aggregate of the steel slag is not clean.

Table 4. Performance index of steel slag.

Technical Indicators		Apparent Relative Density (g/cm³)	Gross Volume Relative Density (g/cm³)	Water Absorption (%)	Water Swelling Rate (%)	Crush Value (%)	Los Angeles Abrasion Value (%)
Particle size/mm	0–3	2.875	2.713	1.0	0.6	-	-
	3–5	3.726	3.511	1.6	1.1	-	15.4
	5–10	3.656	3.449	2.3	1.2	11.2	14.2
	10–30	3.445	3.247	2.6	0.9	13.9	20.1
Standard		≥2.5	-	≤3	≤2	≤22	≤26
Test method		T0304-2005	T0304-2005	T0304-2005	T0348-2005	T0316-2005	T0317-2005

2.2. Microscopic Characterization

2.2.1. X-ray Diffraction (XRD) and X-ray Fluorescence (XRF) Tests

As the main by-product of the steel smelting industry, steel slag has a complex mineral composition and a wide variety of elements. Different mineral elements have different activities and different thermal conductivity. To visually distinguish the difference in mineral composition between the steel slag and limestone, XRD and XRF tests were performed to provide a theoretical basis for the thermal conductivity of the steel slag.

2.2.2. Scanning Electron Microscope (SEM) Test

Due to the formation process of steel slag, its structure is not dense. On the one hand, the porous structure affects the water absorption of the material and the amount of asphalt, as well as the thermal conductivity of the steel slag. To further clarify its surface structure, the microscopic morphology of the steel slag surface was characterized by SEM.

2.3. Design of SSAM

2.3.1. Target Grading

The principle of mineral grading design is to meet the road performance; under this premise, the material composition can be changed to meet the requirements of thermal conductivity. A good asphalt mixture gradation should ensure a reasonable proportion of coarse and fine aggregates and can form a good skeleton structure after mixing with asphalt to generate sufficient strength to resist external deformation. Since the mix proportion design of the SSAM is based on the target gradation of raw materials, the selection of the target gradation and the performance verification are the preconditions for the mix proportion design of the SSAM. According to the results of sieving of each fraction of aggregate in Table 5, an AC-13 (asphalt mixture with a nominal maximum aggregate size of 13 mm) gradation curve close to the center line of gradation was preliminarily selected. The dosages of filler were 0–3, 3–5, 5–10, and 10–15 mm, with aggregate of 6, 31, 11, 19, and 33%, respectively. The Marshall specimen indices of the AC-13 asphalt mixture are shown in Table 6; the calculated optimal asphalt aggregate ratio is 4.9%. The performance of the AC-13 asphalt mixture with the optimal asphalt aggregate ratio is shown in Table 7. The test results meet the requirements of the specification [28].

Table 5. Mineral material and target gradation.

Mesh Size (mm)	16	13.2	9.5	4.75	2.36	1.18	0.6	0.3	0.15	0.075	Content (%)
Filler	100	93.3	25.3	0.6	0.4	0.4	0.4	0.4	0.4	0.4	6
0–3	100	100	92.6	2.7	0.4	0.4	0.3	0.3	0.3	0.2	31
3–5	100	100	100	79	2.6	1.2	0.9	0.8	0.7	0.6	11
5–10	100	100	100	100	86.3	62.7	42.7	20.9	11.4	8.5	19
10–15	100	100	100	100	100	100	100	100	95.9	55.3	33
Upper grading limit (%)	100	100	85	68	50	38	28	20	15	8	/
Median value of grading (%)	100	95	76.5	53	37	26.5	19	13.5	10	6	/
Lower grading limit (%)	100	90	68	38	24	15	10	7	5	4	/
Target grading (%)	100.0	97.8	73.9	46.4	33.2	25.8	19.5	12.8	9.6	6.2	/

Table 6. Marshall specimen indices of AC-13 under target grading.

Asphalt Aggregate Ratio (%)	Gross Bulk Density (g/cm^3)	Void Ratio (%)	VMA (%)	VFA (%)	Stability (kN)	Flow (mm)
3.5	2.375	9.2	16.9	42.2	10.01	2.67
4.0	2.398	6.8	15.7	53.5	10.20	3.02
4.5	2.413	5.2	15.0	63.4	10.97	3.39
5.0	2.425	3.7	14.6	73.2	11.73	3.78
5.5	2.412	3.4	15.1	76.7	9.82	4.52
Test method	T0705-2011	T0705-2011	T0705-2011	T0705-2011	T0709-2011	T0709-2011

Table 7. Performance of AC-13 with the optimal asphalt aggregate ratio.

Performance	Dynamic Stability (Times/mm)	Flexural Tensile Strain (με)	Residual Stability (%)	Freeze–Thaw Splitting Strength Ratio (%)
Test result	3947	5109	96.7	91.5
Standard	≥2800	≥2500	≥80	≥75
Test method	T0719-2011	T0715-2011	T0709-2011	T0729-2011

2.3.2. Steel Slag Particle Size and Replacement Principle

1. Steel slag size selection

The optional 0–3, 3–5, and 5–10 mm steel slags (numbered S-1, S-2, and S-3, respectively) were screened. The sieving results were compared with the sieving results of various fractions of limestone aggregates and used as the basis for the selection of steel slag particle size. The sieving curves are shown in Figure 1. The pass rates of S-1 and S-2 are similar to the fractions of 0–3 and 3–5 mm, respectively, while the pass rate of S-3 is quite different from that of limestone of 5–10 mm. Therefore, it may be easier to use S-1 to replace 0–3 mm limestone aggregate or S-2 to replace 3–5 mm limestone aggregate to prepare SSAM with a gradation close to the target gradation. As S-1 may be impure, S-2 was used to replace the 3–5 mm aggregate. The replacement ratios are 20%, 40%, 60%, 80%, and 100%. The aggregate proportions of different SSAMs are shown in Table 8, and the final gradation curves are shown in Figure 2.

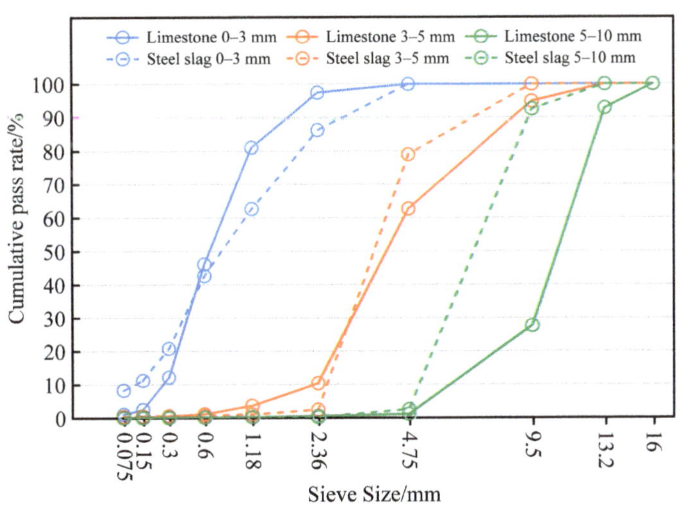

Figure 1. Screening results of steel slag and limestone.

Table 8. Material proportions of SSAM (volume fraction).

Replacement Rate	The Proportion of Various Materials (%)					
	Filler	0–3 mm	3–5 mm	S-2 (Steel Slag)	5–10 mm	10–15 mm
20%	6	31	8.8	2.2	19	33
40%	6	31	6.6	4.4	19	33
60%	6	31	4.4	6.6	19	33
80%	6	31	2.2	8.8	19	33
100%	6	31	0	11	19	33

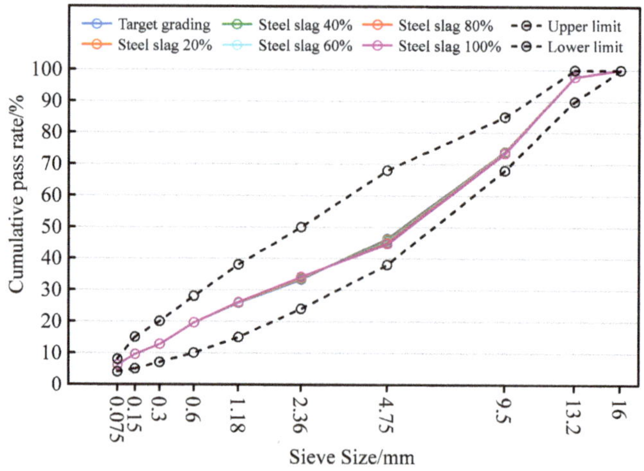

Figure 2. Gradation curves of SSAMs.

2. Replacement principle

The density and water absorption of steel slag are 20% higher than that of natural aggregates, so the design method of mixtures dominated by quality control is not suitable for the gradation design of SSAM. In this study, based on the principle of equal volume replacement, the design of SSAM was conducted by the method of density conversion [29]. After the steel slag replaces the aggregate, the aggregate volume that passes through each sieve hole remains the same. The volume of steel slag and the mass of steel slag were converted by using the relative density of the gross volume of aggregate as a medium. Density conversion is performed according to Equation (1).

$$P_s = P_i \cdot \eta \cdot \frac{\rho_s}{\rho_i} \quad (1)$$

where P_s is the quality of steel slag to be used, P_i is the total weight of a certain type of limestone aggregate to be replaced, η is the volume percentage of aggregate replaced by steel slag, ρ_s is the gross volume relative density of steel slag, and ρ_i is the gross volume relative density of the limestone aggregate to be replaced.

2.4. Determination of Thermal Coefficient of SSAMs

The thermal coefficient of the steel slag mixture was determined concerning the standard "Refractory materials—Determination of thermal conductivity—Hot-wire method" (GB/T5990-2006) [30]. The schematic diagram of the principle of measuring the thermal coefficient by the heating wire method is shown in Figure 3a. The specimens used for the determination of thermal coefficient were small Marshall specimens formed by standard compaction methods. The size of the specimen is φ 101.6 × 63.5 mm^3. The treated specimen is shown in Figure 3b. The H hole is located at the center of the upper surface of the Marshall specimen. The hole diameter is 6 mm and the depth is 50 mm for embedding the heating wire. Holes P1, P2, and P3 are used to embed temperature sensors. The three holes are arranged equidistant from the center hole H, and the spacing meets the requirements of 20 ± 5 mm. After the hole is drilled, the debris inside the specimen should be cleaned. Subsequently, the embedding of the heating wire and the temperature sensor is performed, as shown in Figure 3c. After the heating wire and the temperature sensor are inserted into the corresponding holes, they are sealed with thermally conductive silicone. The heating wire used is a Ni–Cr alloy with a diameter of 6 mm, a length of 50 mm, and a heating power of 60 W. It has the characteristics of fast heating and good heat resistance. The temperature sensor is a PT100 platinum resistance thermometer with a diameter of 4 mm and a length of 30 mm. The working temperature of the temperature sensor is −50–200 °C, and the temperature error is 0.1 °C.

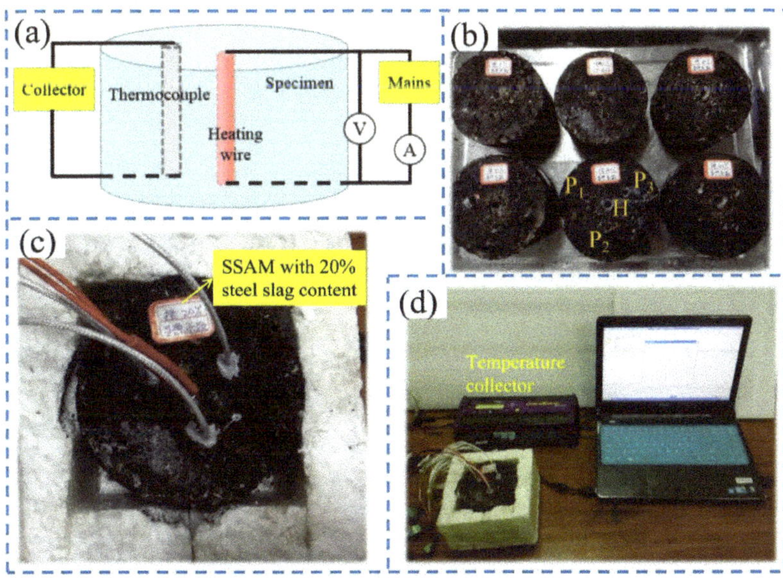

Figure 3. The principle and preparation of thermal coefficient measurement. (**a**) Test principle, (**b**) drilled specimen, (**c**) heat preservation of specimen, (**d**) test system.

Figure 3d shows the framework of the thermal coefficient test system, of which the main parts are a computer, a data acquisition instrument, a heating wire, and a temperature sensor. When the ambient temperature fluctuation during the test is lower than 0.3 °C within 2 min, the power supply is started to heat the heating wire. The data acquisition instrument records data every 5 s. In the processes of power-on and heating, a polystyrene foam board is used for thermal insulation of the specimen. In this way, the heat exchange between the side surface of the specimen and the outside air is prevented, thereby reducing the test error. The temperature rise curve of the asphalt mixture under each steel slag content can be drawn, and the thermal coefficient of the SSAM can be obtained by substituting the temperature data into Equation (2).

$$\lambda = \frac{UI}{4\pi L} \cdot \frac{-E_i(\frac{-r^2}{4\alpha t})}{\Delta\theta_2 - \Delta\theta_1} \quad (2)$$

where λ is the thermal coefficient of the specimen (W/(m·°C)), U is the voltage of the heating wire (V), I is the current of the heating wire (A), L is the length of the heating wire embedded in the specimen (m), r is the distance between the thermocouple and the heating wire (m), α is the thermal diffusivity (m^2/s), t_1 and t_2 are the measurement time in the stable state of temperature rise during the measurement process (s), $\Delta\theta_1$ and $\Delta\theta_2$ are the temperature rise at the measurement time t_1, t_2 (°C), and $-E_i(\frac{-r^2}{4\alpha t})$ is the exponential integral of $\int_x^u \frac{-e^u du}{u}$, and its value can be found through the ratio table of $\frac{\Delta\theta_2}{\Delta\theta_1}$ [30].

2.5. Volume Indices of SSAM

Due to the porosity of steel slag, the diversity of chemical elements, and the complexity of composition, the content of steel slag affects the performance of asphalt mixtures. Six substitution levels were designed to replace 0, 20, 40, 60, 80, and 100% of aggregates 3–5 by at equal volume. Under the condition that the optimal amount of asphalt remains unchanged, Marshall specimens of SSAMs were prepared.

Due to the significant density difference between steel slag and natural aggregate of the same size, replacing limestone with an equal volume of steel slag increases the quality of the asphalt mixture. The asphalt aggregate ratio is expressed as the percentage of asphalt mass to mineral aggregate. Due to the addition of steel slag, the original asphalt aggregate ratio has changed under the condition of the same amount of asphalt. Therefore, it is necessary to redefine the asphalt aggregate ratio after adding steel slag. In this paper, the asphalt aggregate ratio in the SSAM is defined as the nominal asphalt aggregate ratio, which reflects the actual amount of asphalt in the mixture. The asphalt aggregate ratio, the nominal asphalt aggregate ratio, and the conversion relationship are calculated according to Equations (3)–(5) respectively.

$$P_a = \frac{m_a}{M} \times 100\% \tag{3}$$

$$P'_a = \frac{m_a}{M + m_i \cdot \eta \cdot (\frac{\rho_s}{\rho_i} - 1)} \times 100\% \tag{4}$$

$$P'_a = P_a \cdot \frac{1}{1 + P_i \cdot \eta \cdot (\frac{\rho_s}{\rho_i} - 1)} \times 100\% \tag{5}$$

where P_a is the asphalt aggregate ratio of the asphalt mixture (%), P'_a is the nominal asphalt aggregate ratio of the SSAM (%), m_a is the asphalt mass (g), M is the total mass of aggregate without adding steel slag (g), m_i is the aggregate mass replaced by steel slag (g), η is the volume percentage of aggregate replaced by steel slag (%), ρ_s is the gross volume relative density of steel slag (g/cm^3), ρ_s is the relative density of the gross volume of the aggregate to be replaced (g/cm^3), and P_i is the total weight of a certain type of limestone aggregate to be replaced (kg).

2.6. Road Performance Test of SSAM

For the SSAM to have the potential to melt ice and snow, it should have good road performance as well as good thermal conductivity. Steel slag was added to the asphalt mixture to replace the aggregate, and the SSAM was prepared under the condition that the optimal amount of asphalt remained unchanged. Through high-temperature rutting tests, low-temperature trabecular bending tests, and freeze–thaw splitting tests, the effects of different amounts of steel slag on the road performance of asphalt mixture were studied.

2.6.1. Rutting Test

At present, the test method used to evaluate the high-temperature stability of asphalt mixture is mainly the rutting test. The rutting test is used to evaluate the high-temperature deformation resistance of the asphalt mixture according to the standard test method (T0719-2011) [27]. The testing effect is intuitive and clear. The size of the rutting plate specimen is 300 × 300 × 50 mm^3. The specimens were formed by the wheel rolling method (T0703-2011) [27], and the compaction temperature was 160 °C. The temperature during the rutting test was 60 °C and the rubber wheel pressure was 0.7 MPa. The correlation between high-temperature performance and rutting depth is characterized by dynamic stability. Dynamic stability is the ratio of the number of rolling runs to the depth of the rut. The greater the dynamic stability, the stronger the deformation resistance of the specimen, and the better the high-temperature stability of the asphalt mixture.

2.6.2. Low-Temperature Bending Test

When the temperature is lower in winter, the volume of the asphalt surface tends to shrink. However, the asphalt surface layer cannot shrink normally under the restraint of the surrounding materials, so thermal stress is generated inside the structure. When the generated temperature stress is greater than the allowable tensile stress of the asphalt mixture, the asphalt mixture is pulled and cracked, resulting in cracks in the asphalt pavement. Therefore, the asphalt mixture must have good resistance to low temperature cracking to ensure the normal use of the road surface. The low-temperature deformation resistance of asphalt mixture is usually evaluated by low-temperature bending tests according to the standard test method (T0715-2011) [27]. The specimens were formed in the same way as in Section 2.6.1, but the specimens were cut into beams after forming. The evaluation index is the maximum bending tensile strain of the specimen at low-temperature failure. The larger the bending tensile strain, the better the low-temperature flexibility of the asphalt mixture; that is, the better the crack resistance.

2.6.3. Retained Marshall Stability Test and Freeze–Thaw Split Test

The insufficient moisture resistance of asphalt mixtures is manifested in the reduction of the cohesion of asphalt mixtures under the action of water. As a result, the asphalt peels off the surface of the aggregate particles, and the pavement has potholes. To ensure the service level and service performance of the road, the water damage resistance of the asphalt mixture should be improved. The moisture resistance evaluation methods include the retained Marshall stability test and freeze–thaw split test. The corresponding indicators are the residual stability ratio and the freeze–thaw splitting strength ratio. These two methods were used according to specifications (T0709-2011 and T0729-2011) [27] to evaluate the moisture resistance of the SSAM. The specimens were formed by the compaction method (T0702-2011) [27], and the compaction temperature was 160 °C.

3. Results and Discussion

3.1. Microscopic Characterization of Thermal Conductive Phase

3.1.1. Mineral Composition

It can be seen from the XRD pattern of the steel slag in Figure 4 that the mineral components in the steel slag include iron-containing oxides (FeO), magnesium-containing oxides (MgO), aluminum-containing oxides (AlO), etc. The mineral component of limestone is relatively simple; most of the peaks point to calcite ($CaCO_3$). The main mineral components of steel slag and limestone are shown in Table 9. It is worth noting that since the XRD analysis software does not show the peak positions of SiO_2 and MnO, the mineral composition displayed by the XRD test is somewhat different from the XRF test.

Studies have shown [31,32] that the heating rate of minerals is related to the activity of their composition. The better the activity of the material, the faster the heating rate; that is, the stronger the thermal conductivity. According to the different heating rates of mineral components, the components can be classified as more active (Fe_3O_4), active (Fe_2O_3, FeS), less active (MgO, Al_2O_3), and inactive (SiO_2, CaO). Minerals with higher iron content are more active. The classification results of the ingredients are shown in Figure 5.

There are many kinds of minerals in steel slag, notably the presence of iron-containing active minerals, which not only increases the activity of steel slag but also greatly improves the thermal conductivity of steel slag itself. On the contrary, limestone only contains calcite, an inactive mineral, which has poor thermal conductivity.

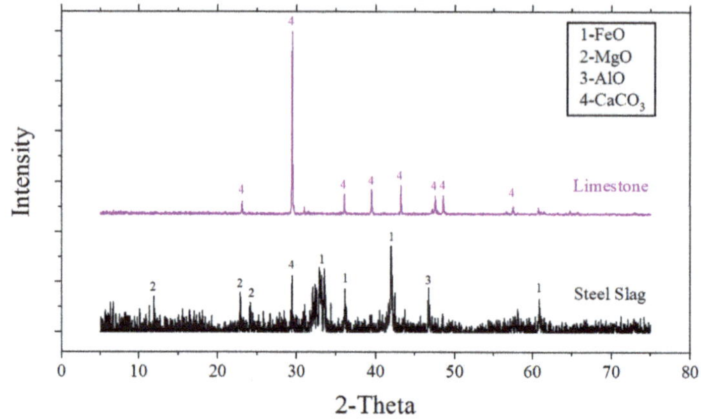

Figure 4. XRD spectra of steel slag and limestone.

Table 9. Main mineral composition of steel slag and limestone.

Materials	Mineral Composition (%)					
	CaO	Fe_2O_3	SiO_2	MgO	MnO	Al_2O_3
Steel slag	37.15	31.28	14.40	5.38	4.28	2.77
Limestone	54.37	0.23	0.74	0.86	0.01	0.37

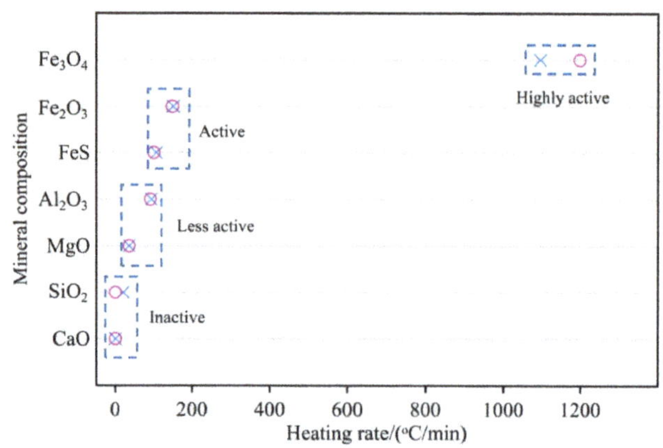

Figure 5. Heating activity of some mineral components [31,32].

3.1.2. Surface Topography

Steel slag is a typical porous material with higher porosity than limestone, so steel slag has a stronger adsorption capacity for water and asphalt. In addition, the thermal conductivity of the solid components in the steel slag is significantly higher than that of the air in the pores. Therefore, the larger the slag voids, the poorer the thermal conductivity may be. Figure 6 shows the SEM images of steel slag at 5000× and 40,000× magnification. In Figure 6a,b, large and small protrusions can be seen on the surface of the steel slag, which makes the surface very rough. In Figure 6b, the micropores on the surface of the steel slag can be clearly seen. The protrusions and micropores of steel slag may increase the porosity of SSAM [29], which may affect the overall thermal conductivity of the SSAM.

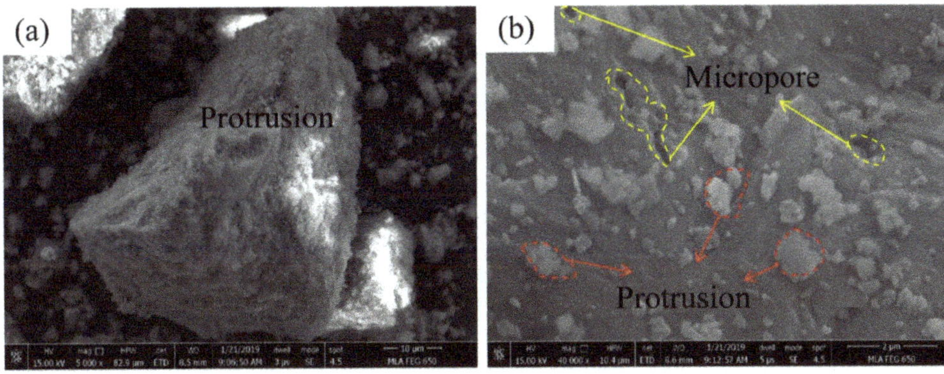

Figure 6. Microscopic surface morphology of steel slag. (**a**) 5000×; (**b**) 40,000×.

3.2. *Thermal Coefficient of SSAM*

3.2.1. Measured Thermal Coefficient

During the heating process, the current in the heating wire is constant, and the heating power remains unchanged. Therefore, the rate of change of temperature rise measured by the thermocouple can directly reflect the thermal conductivity of the specimen. The temperature rise curves of asphalt mixtures with different steel slag contents are shown in Figure 7. During a heating period of 30–75 s, the mixture between the temperature sensor and the heating wire is preheating. The heat transfer to the sensor is lower and the temperature rise is smaller. The recorded data are the ambient temperature in the initial state. With the prolongation of heating time, the mixture gradually began to heat up. The heating wire temperature is transmitted to the sensor through the mixture. Elevated temperature and time are not linear in the transition phase due to heat transfer instabilities. When heated to 90 s, the heat transfer is stable, the mixture heats up sharply, and the temperature rise is almost linearly related to time. This obvious phenomenon can be observed in the temperature rise curves of asphalt mixtures under all slag contents. Theoretically, the slope of the temperature rise curve can reflect the heating rate of the material. Although the data collected and recorded by the three temperature sensors in the same specimen are slightly different, the slopes of the three curves are the same. This indicates that the distribution of steel slag in the asphalt mixture is relatively uniform, and the results of the test method are reliable.

The thermal coefficient of the asphalt mixture under different steel slag contents is shown in Figure 8. With the increase in steel slag content, the thermal coefficient of asphalt mixtures first increases and then decreases. When the content of steel slag is about 60%, the thermal coefficient of SSAM is the largest, which is about 120.3% of the thermal coefficient without steel slag. The thermal coefficient of asphalt mixture with 20% and 40% steel slag content also increased to a certain extent compared with 0% content, increasing by 11.4% and 14.6%, respectively. When the content of steel slag exceeds 60%, the thermal coefficient obviously decreases. When the content is 100%, the steel slag completely replaces the aggregate of 3–5 mm, and the thermal coefficient of the asphalt mixture is even lower than that of the original mixture (11.1%). This phenomenon occurs because steel slag is a typical porous material, and its porosity is significantly higher than that of the limestone aggregate. As the steel slag content increases, the porosity of the SSAM increases, which is verified in Section 3.3. Solids conduct heat better than air. Therefore, the higher slag content hinders the improvement of the thermal conductivity of SSAM. Although more slag means more iron-containing oxides in SSAM, when the slag content is high, the effect of thermal conductivity reduction caused by increased porosity may be more prominent. When the content of steel slag is moderate, the highly active metal elements in the steel slag may play a major role relative to the porosity, so the thermal coefficient of the SSAM is higher than that of the asphalt mixture without steel slag.

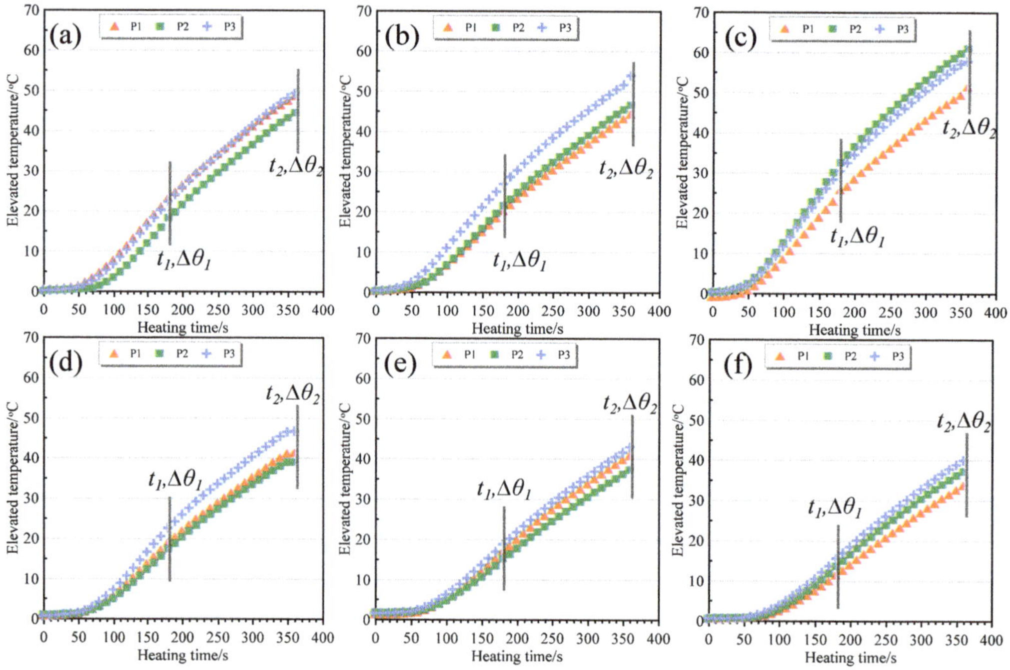

Figure 7. Heating curve. (**a**) 0 content; (**b**) 20% content; (**c**) 40% content; (**d**) 60% content; (**e**) 80% content; (**f**) 100% content.

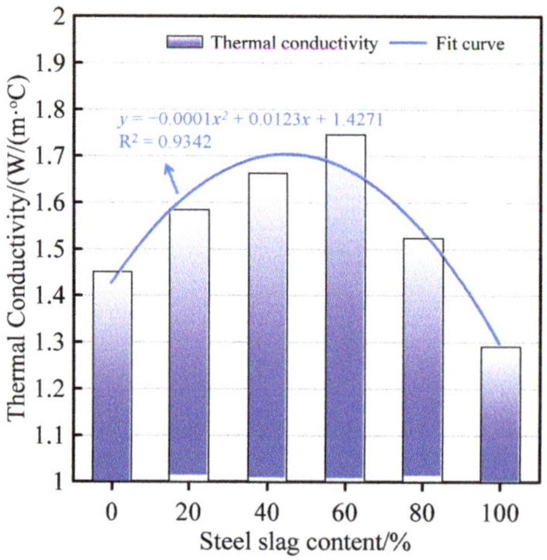

Figure 8. Thermal coefficient of asphalt mixtures with different slag content.

3.2.2. Calculated Thermal Coefficient

There are many test methods for the thermal coefficient of asphalt mixture, and different test methods have different test results due to different test principles and test conditions. Theoretical calculations determine the theoretical reference value according to the material composition characteristics. The theoretical value and the actual measurement value can be mutually verified to verify the feasibility of the test method to a certain extent.

In 1972, Williamson [33] proposed the theoretical calculation Equation (6) of the thermal coefficient prediction model of asphalt mixture based on the research on the thermal properties of asphalt mixture.

$$\lambda_{ac} = (\lambda_b)^{V_b} \cdot (\lambda_m)^{V_m} \cdot (\lambda_a)^{V_a} \cdot (\lambda_w)^{V_w} \quad (6)$$

where λ_{ac} is the thermal coefficient of the asphalt mixture; λ_b, λ_m, λ_a, and λ_w are the thermal coefficient of asphalt, aggregate, air, and water, respectively; and V_b, V_m, V_a, and V_w are the volume percentages of asphalt, aggregate, air, and water in the asphalt mixture, respectively.

According to the test data in Table 10, when the content of steel slag is 0%, the density of the asphalt mixture is 2.420 g/cm^3, and the porosity is 3.71%. Combined with the density of each raw material in Tables 1 and 2, it can be calculated that the volume percentages of asphalt, mineral material, and water in the asphalt mixture are 11.11, 85.12, and 0.06%, respectively. The thermal coefficients of asphalt, limestone aggregate, air, and water were taken as 0.7, 2.04, 0.024, and 0.6 W/(m·°C), respectively. By substituting the above data into Equation (6), λ_{ac} can be obtained as 1.3821 W/(m·°C).

Table 10. Test results for Marshall specimens.

Asphalt Aggregate Ratio (%)	Steel Slag Content (%)	Nominal Asphalt Aggregate Ratio (%)	Gross Volume Relative Density	Water Absorption (%)	Void Ratio (%)
4.90	0	4.90	2.420	0.4	3.71
	20	4.87	2.447	0.6	3.70
	40	4.83	2.460	0.8	3.78
	60	4.80	2.469	1.0	3.82
	80	4.77	2.478	1.1	4.13
	100	4.73	2.489	1.3	4.37

The theoretically calculated thermal coefficient is 1.3821 W/cm^3, which is only 4.78% different from the measured value of 1.4515 W/cm^3 measured by the parallel heating wire method. The thermal coefficient of asphalt mixtures measured by the heating wire method is relatively reliable.

3.3. Volume Indices of SSAM

The test results of the SSAM are shown in Table 10. With the increase of steel slag content, the nominal asphalt aggregate ratio and gross volume relative density increase, because the density of steel slag aggregate is higher than that of limestone aggregate. At the same time, the water absorption and porosity of SSAM also increase with an increase in steel slag content. Because the gradation curves of each SSAM are similar, the reason for the increase in porosity may be the increase in the porosity of the steel slag itself. In addition, the roughness of the steel slag surface reduces the workability of the SSAM, which may also be a factor affecting the porosity of the SSAM [29].

3.4. Road Performance Verification of SSAM

3.4.1. High-Temperature Performance

The test results of the dynamic stability of the asphalt mixture with different steel slag contents are shown in Figure 9 below. The dynamic stability of asphalt mixtures first increases and then decreases with the increase of steel slag content. When the steel slag completely replaces the aggregate, the dynamic stability is even smaller than that without steel slag. When the slag content increased from 0% to 60%, the dynamic stability value increased from 3947 to 5732, an increase of 45.2%. An appropriate amount of steel slag can significantly increase the high-temperature deformation resistance of the asphalt mixture, which is due to the large specific gravity and high strength of the steel slag itself. The close contact between the steel slag and the aggregate enhances the integrity of the structure, improving the resistance to high-temperature deformation. However, with the further increase in steel slag content, the pores of the mixture increase. Pore adsorption of asphalt reduces the free asphalt content in the mixture and reduces the adhesion between aggregates. The kneading effect at high-temperature conditions causes the low-adhesion aggregates to loosen and fall off, which eventually leads to a decrease in the strength and high-temperature stability of the asphalt mixture.

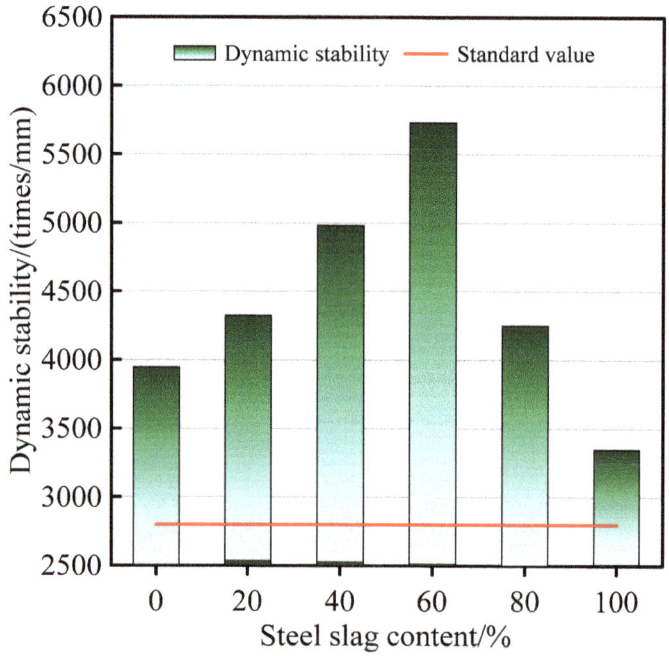

Figure 9. High-temperature performance of SSAM.

3.4.2. Low-Temperature Performance

The bending test results of asphalt mixture trabeculae with different steel slag contents are shown in Figure 10 below. With the increase in steel slag content, the flexural and tensile strain of asphalt mixture at low-temperature failure decreases linearly, and the range of decrease is large. When the content exceeds 80%, it does not meet the requirements of the specification, indicating that the addition of steel slag has a great influence on the low-temperature crack resistance of the asphalt mixture. This is because the steel slag has more pores and a larger amount of adsorbed asphalt per unit volume, which reduces the adhesion of asphalt to aggregates. The cohesion between the aggregate particles is reduced, and it is not sufficient to resist the shrinkage deformation caused by the temperature stress. In addition, with the increase in steel slag content, the porosity of asphalt mixtures increases. The steel slag adsorbs the asphalt, resulting in a significant stress relaxation effect near the steel slag. However, the asphalt content in the SSAM is fixed, which makes the asphalt content in some non-steel slag areas insufficient. Overall, the SSAM lacks the cementation of asphalt, making it more susceptible to cracking from thermal stress. In general, steel slag reduces the low-temperature performance of the asphalt mixture and the replacement ratio of steel slag should not exceed 60 %.

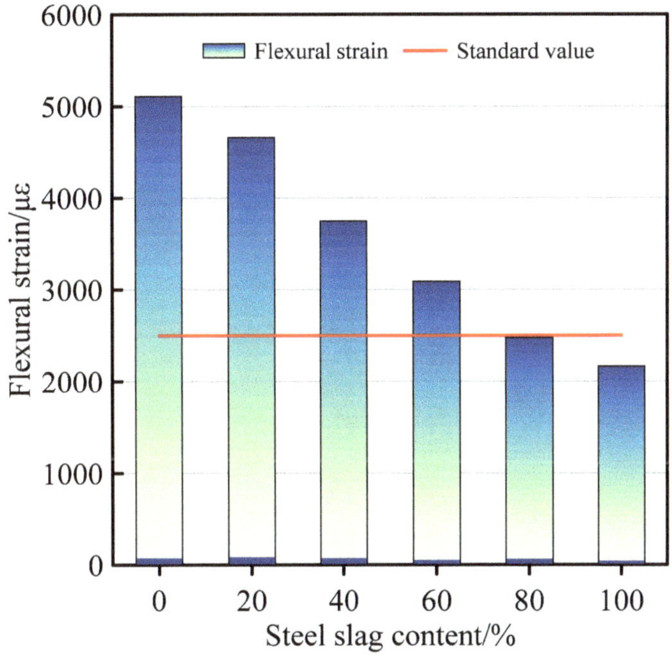

Figure 10. Low-temperature performance of SSAM.

3.4.3. Moisture Resistance

As shown in Figure 11, as the steel slag increases, the freeze–thaw splitting strength ratio and residual stability of SSAMs decreases. The incorporation of steel slag reduces the moisture resistance of the asphalt mixture. For the freeze–thaw splitting strength index, the addition of steel slag reduces the water damage resistance of the asphalt mixture, resulting in a maximum reduction of the freeze–thaw splitting strength ratio by 9%. However, the freeze–thaw split index still meets the specification value of 75%. For the residual stability index, when the slag content is greater than 80%, the residual stability drops sharply. The maximum decrease in residual stability is 17.5%, which does not meet the specification value of 80%. The steel slag affects the moisture resistance of the asphalt mixture because the steel slag absorbs a large amount of asphalt, which reduces the free asphalt in the asphalt mixture. The asphalt film on the aggregate surface becomes thinner. Water has a stronger adsorption force to the aggregate, causing the water to pass through the asphalt membrane to separate the aggregate from the asphalt. As a result, the asphalt peels off the surface of the aggregate, the mixture becomes loose, and the strength is reduced. In addition, with the increase in the amount of steel slag, the porosity of the asphalt mixture increases, so the moisture can more easily enter the interior of the asphalt mixture structure. Under the action of hydrodynamic pressure, the asphalt migrates or even peels off from the aggregate surface, which reduces the moisture resistance of the asphalt mixture. In addition, incomplete slag aging may also be a factor in reducing moisture resistance [34]. Similar conclusions are also reflected in some literature [35,36]. Of course, some scholars have pointed out that with the increase in steel slag content, the moisture resistance of SSAM increases [29,37,38]. This is because the adhesion of steel slag to asphalt is higher than that of ordinary aggregate. Therefore, in conclusion, the good adhesion performance of steel slag and asphalt has a dual effect, which may lead to increased asphalt content near the steel slag and decreased asphalt content in the non-steel slag area. The effect of steel slag on the moisture resistance of SSAM is still open for discussion.

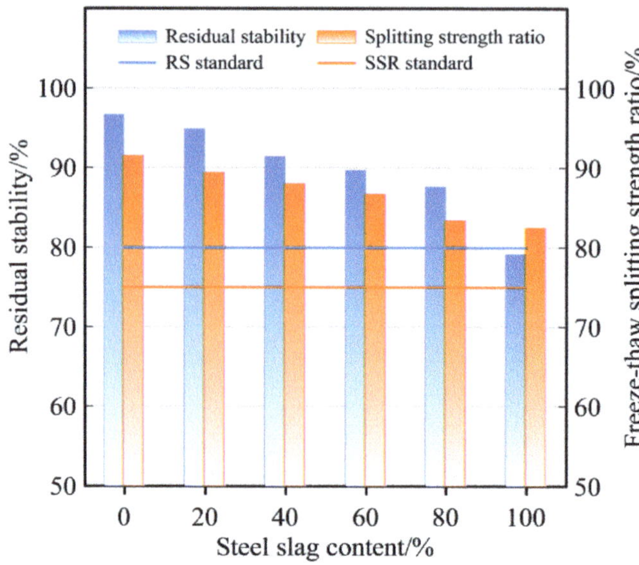

Figure 11. Moisture resistance of SSAM.

In terms of road performance, the addition of steel slag can significantly improve the high-temperature rutting resistance of asphalt mixtures, but it can also reduce the low-temperature crack resistance and moisture resistance in some cases. Combined with the thermal conductivity and road performance of asphalt mixture under different steel slag content, when the steel slag content is 60%, the asphalt mixture has the best high-temperature stability and the highest thermal conductivity. Under these conditions, the low-temperature crack resistance and moisture resistance meet the requirements of the technical specification [28]. Therefore, the optimum content of steel slag content should not exceed 60% (6.6% of the total mixture).

4. Conclusions

To better utilize the functional properties of steel slag in asphalt mixtures, such as ice and snow melting, self-healing, etc., the microscopic properties and thermal conductivity of steel slag and SSAM were studied in this paper. Further, to ensure the SSAM meets the engineering requirements, the basic road performance of SSAMs with different thermal conductivity is verified. The main conclusions are as follows:

(1) The XRD and XRF tests of steel slag showed that, compared with limestone, iron-containing active oxides are the main factor driving the heat conduction of steel slag. The SEM test showed that the surface of the steel slag was rough and microporous, which may be detrimental to the thermal conductivity of the SSAM.

(2) The thermal conductivity of SSAM first increases and then decreases with the content of steel slag. When the content of slag is less than 60%, the beneficial effect of iron-containing oxides on thermal conductivity may be dominant. When the content is greater than 60%, the main factor affecting the thermal conductivity of steel slag may be the increased voids in the mixture. The porosity of SSAM increases with the increase of steel slag content, which supports this conjecture.

(3) When the steel slag content is 60%, the highest thermal coefficient of 1.746 W/(m·°C) is reached. The difference between the test value and the theoretical value is 4.78%, which verifies the reliability of the test result.

(4) The road performance tests show that there is an optimal content to make the high-temperature performance optimal, while the low-temperature performance and moisture resistance decrease with increasing slag content.
(5) After considering the thermal conductivity and road performance of SSAM for functional applications, the optimum amount of steel slag should not exceed 60% of the aggregate volume of 3–5 mm (6.6% of the total SSAM). In this paper, a preliminary evaluation of the thermal conductivity of SSAM is presented, and the long-term performance of the mixture needs to be further studied.

Author Contributions: Conceptualization, Y.C.; methodology, Y.C.; validation, H.L.; formal analysis, Y.C.; investigation, Y.C., J.L.; resources, H.L., F.Z.; data curation, H.L., F.Z.; writing—original draft preparation, Y.C.; writing—review and editing, Y.C., J.L.; visualization, F.Z., J.L.; supervision, Z.L.; project administration, A.S.; funding acquisition, A.S. All authors have read and agreed to the published version of the manuscript.

Funding: This research was funded by the National Key R&D Program of China (Grant No. 2021YFB1600200) and Scientific Innovation Practice Project of Postgraduates of Chang'an University (300103722016).

Institutional Review Board Statement: Not applicable.

Informed Consent Statement: Not applicable.

Data Availability Statement: All data used during the study appear in the published article.

Conflicts of Interest: The authors declare no conflict of interest.

References

1. Pasetto, M.; Baliello, A.; Giacomello, G.; Pasquini, E. Sustainable solutions for road pavements: A multi-scale characterization of warm mix asphalts containing steel slags. *J. Clean. Prod.* **2017**, *166*, 835–843. [CrossRef]
2. Li, L.; Ling, T.; Pan, S. Environmental benefit assessment of steel slag utilization and carbonation: A systematic review. *Sci. Total Environ.* **2022**, *806*, 150280. [CrossRef]
3. Dai, J.; Ma, F.; Fu, Z.; Li, C.; Jia, M.; Shi, K.; Wen, Y.; Wang, W. Applicability assessment of stearic acid/palmitic acid binary eutectic phase change material in cooling pavement. *Renew. Energy* **2021**, *175*, 748–759. [CrossRef]
4. Cao, Y.; Sha, A.; Liu, Z.; Li, J.; Jiang, W. Energy output of piezoelectric transducers and pavements under simulated traffic load. *J. Clean. Prod.* **2021**, *279*, 123508. [CrossRef]
5. Mo, L.; Lin, S.; Meng, X.; Qu, L.; Chang, W.; Xiao, Y. Volume expansion characteristics and crack simulation of steel slag. *China J. Highw. Transp.* **2021**, *34*, 180–189. [CrossRef]
6. Wang, C.; Wang, M.; Chen, Q.; Zhang, L. Basic performance and asphalt smoke absorption effect of environment-friendly asphalt to improve pavement construction environment. *J. Clean. Prod.* **2022**, *333*, 130142. [CrossRef]
7. Kambole, C.; Paige-Green, P.; Kupolati, W.; Ndambuki, J.; Adeboje, A. Basic oxygen furnace slag for road pavements: A review of material characteristics and performance for effective utilisation in southern Africa. *Constr. Build. Mater.* **2017**, *148*, 618–631. [CrossRef]
8. Cao, Y.; Zhang, F.; Sha, A.; Liu, Z.; Hao, Y.; Hao, Y. Energy conversion models and characteristics under various inner connections of a novel packaged piezoelectric transducer for pavements. *Energy Convers. Manag.* **2021**, *245*, 114563. [CrossRef]
9. Kavussi, A.; Qazizadeh, M. Fatigue characterization of asphalt mixes containing electric arc furnace (EAF) steel slag subjected to long term aging. *Constr. Build. Mater.* **2014**, *72*, 158–166. [CrossRef]
10. Cao, J. Effect of surface modification of steel slag on fatigue durability of asphalt mixture. *New Build. Mater.* **2020**, *47*, 32–35+44.
11. Kara, M.; Günay, E.; Kavakli, B.; Tayfur, S.; Eren, K.; Karadag, G. The use of steel slag in asphaltic mixture. *Key Eng. Mater.* **2004**, *264-268*, 2493–2496. [CrossRef]
12. Zhang, C.; Wang, C.; Li, S.; Ding, W.; Tang, X. Research on water stability of steel slag asphalt mixture. *Bull. Chin. Ceram. Soc.* **2021**, *40*, 207–214. [CrossRef]
13. Masoudi, S.; Abtahi, S.; Goli, A. Evaluation of electric arc furnace steel slag coarse aggregate in warm mix asphalt subjected to long-term aging. *Constr. Build. Mater.* **2017**, *135*, 260–266. [CrossRef]
14. Ahmedzade, P.; Sengoz, B. Evaluation of steel slag coarse aggregate in hot mix asphalt concrete. *J. Hazard. Mater.* **2009**, *165*, 300–305. [CrossRef] [PubMed]
15. Zeng, G.; Liu, H.; Bai, F.; Zhou, P. Research on visco-elastoplastic constitutive model for ac-13 graded asphalt mixture with steel slag. *Bull. Chin. Ceram. Soc.* **2020**, *39*, 4061–4067. [CrossRef]
16. Ameri, M.; Hesami, S.; Goli, H. Laboratory evaluation of warm mix asphalt mixtures containing electric arc furnace (EAF) steel slag. *Constr. Build. Mater.* **2013**, *49*, 611–617. [CrossRef]

17. Chen, Z.; Leng, Z.; Xiao, Y.; Jiang, J.; Jiao, Y.; Wu, S.; Xie, J.; Cai, J. Complete replacement of mineral raw materials in asphalt concrete by steel & iron slags based on cascade utilization strategy. *China J. Highw. Transp.* **2021**, *34*, 190–203. [CrossRef]
18. Javilla, B.; Fang, H.; Mo, L.; Shu, B.; Wu, S. Test evaluation of rutting performance indicators of asphalt mixtures. *Constr. Build. Mater.* **2017**, *155*, 1215–1223. [CrossRef]
19. Polaczyk, P.; Ma, Y.; Xiao, R.; Hu, W.; Jiang, X.; Huang, B. Characterization of aggregate interlocking in hot mix asphalt by mechanistic performance tests. *Road Mater. Pavement Des.* **2021**, *22*, S498–S513. [CrossRef]
20. Huang, S. Compaction characteristic research steel slag drainage asphalt mixture. *Highw. Eng.* **2016**, *41*, 294–298.
21. Martinho, F.; Picado-Santos, L.; Capitão, S. Influence of recycled concrete and steel slag aggregates on warm-mix asphalt properties. *Constr. Build. Mater.* **2018**, *185*, 684–696. [CrossRef]
22. Liu, H.; Zhu, B.; Wei, H.; Chai, C.; Chen, Y. Laboratory evaluation on the performance of porous asphalt mixture with steel slag for seasonal frozen regions. *Sustainability* **2019**, *11*, 6924. [CrossRef]
23. Crisman, B.; Ossich, G.; Lorenzi, L.; Bevilacqua, P.; Roberti, R. A laboratory assessment of the influence of crumb rubber in hot mix asphalt with recycled steel slag. *Sustainability* **2020**, *12*, 8045. [CrossRef]
24. Xiang, L. Study on the preparation and properties of three-phase conductive asphalt mixture. *New Build. Mater.* **2017**, *44*, 137–140.
25. Xiang, Y.; Liu, W.; Zhao, Y.; Zhang, Q.; Zhang, Y. Self-healing Performance of SSAMs by Microwave Heating. *Bull. Chin. Ceram. Soc.* **2022**, *41*, 667–677. [CrossRef]
26. Gao, J.; Sha, A.; Wang, Z.; Tong, Z.; Liu, Z. Utilization of steel slag as aggregate in asphalt mixtures for microwave deicing. *J. Clean. Prod.* **2017**, *152*, 429–442. [CrossRef]
27. *JTG E20-2011*; Standard Test Methods of Bitumen and Bituminous Mixtures for Highway Engineering. China Communications Press Co., Ltd.: Beijing, China, 2011.
28. *JTG F40-2004;* Technical Specifications for Highway Asphalt Pavements. China Communications Press Co., Ltd.: Beijing, China, 2004.
29. Moura, C.; Nascimento, L.; Loureiro, C.; Rodrigues, M.; Oliveira, J.; Silva, H. Viability of using high amounts of steel slag aggregates to improve the circularity and performance of asphalt mixtures. *Appl. Sci.* **2022**, *12*, 490 doiorg/103390/app12010490. [CrossRef]
30. *GB/T 5990-2006*; Refractory Materials—Determination of Thermal Conductivity—Hot-Wire Method. China Quality and Standards Publishing & Media Co., Ltd.: Beijing, China, 2006.
31. Tinga, W. Microwave dielectric constants of metal oxides at high temperature: Part 2. *Electromagn. Energy Rev.* **1989**, *1*, 2–6.
32. Xia, D.; Pickles, C. Applications of microwave energy in extractive metallurgy, a review. *CIM Bull.* **1997**, *90*, 96–107.
33. Williamson, R. Effects of Environment on Pavement Temperatures. In *Proceeding of the 3rd International Conference on Structural Design of Asphalt Pavement, London, UK, 11–15 September 1972;* The National Academies Press: Washington, DC, USA, 1972; pp. 144–158.
34. Qin, L. Influence of ageing effect on steel slag and performance of asphalt concrete volume and its water stabilit. *J. China Foreign Highw.* **2019**, *39*, 264–270. [CrossRef]
35. Yang, Y.; Wu, J.; Zhang, J. Research on mix design and pavement performance of sma-13 asphalt mixture with steel slag. *Road Mach. Constr. Mech.* **2016**, *33*, 36–40.
36. Hadi, G.; Saeid, H.; Mahmoud, A. Laboratory evaluation of damage behavior of warm mix asphalt containing steel slag aggregates. *J. Mater. Civ. Eng.* **2017**, *29*, 04017009. [CrossRef]
37. Gao, Z.; Shen, A.; Zhai, C.; Guo, Y.; Yu, P. Determination of volumetric parameters and impacting mechanism of water stability for steel slag asphalt mixture. *J. Traffic Transp. Eng.* **2018**, *18*, 1–10. [CrossRef]
38. Lyu, Z.; Shen, A.; Li, D.; Guo, Y.; Zhai, C.; Yang, X. Effect of dry–wet and freeze–thaw repeated cycles on water resistance of steel slag asphalt mixture. *Iran. J. Sci. Technol.-Trans. Civ. Eng.* **2021**, *45*, 291–301. [CrossRef]

MDPI
St. Alban-Anlage 66
4052 Basel
Switzerland
www.mdpi.com

Sustainability Editorial Office
E-mail: sustainability@mdpi.com
www.mdpi.com/journal/sustainability

www.ingramcontent.com/pod-product-compliance
Lightning Source LLC
Chambersburg PA
CBHW061611210926
51850CB00009B/2291
* 9 7 8 3 7 2 5 8 0 1 1 4 5 *